SPACE FLIGHT

Space Flight

ARTHUR I. BERMAN

Illustrated by the author except where noted

SCIENCE STUDY SERIES

A DOUBLEDAY ANCHOR ORIGINAL
ANCHOR PRESS/DOUBLEDAY
GARDEN CITY, NEW YORK
1979

Printed in the United States of America
First Edition

Library of Congress Cataloging in Publication Data

Berman, Arthur I.
Space flight.

(Science study series)
Includes index.
1. Space flight. I. Title. II. Series.
TL790.B46 629.4
ISBN: 0-385-02754-0
Library of Congress Catalog Card Number: 74-18841

Acknowledgment

The author thanks the National Aeronautics and Space Administration and the Jet Propulsion Laboratory for furnishing illustrative material, *Scientific American* for permission to reproduce copyrighted drawings, and John Wiley & Sons for releasing portions of *The Physical Principles of Astronautics*. Arthur C. Clarke kindly permitted a paragraph in *Rendezvous with Rama* to be included in the chapter on perturbations. Curtain Call Productions allowed portions of Allan Sherman's "Shine On, Harvey Bloom" to be reproduced in the chapter on lunar launching.

to

Al Baez
Hans Flyger
Niels Meyer
John Philip
Warren Proctor
Bjørn Qvale

in appreciation

Contents

Foreword

This book is for people who love to visit planetariums and love to watch films which describe, with just a bit of poetry and music, how it would feel to encounter closeup friendly visitors from another world. Or watch an expedition from the Moon to Jupiter unfold under the watchful eyes of a computer in search of a conscience. Or discover with hidden grief that life (and lifestyle) long, long ago on a very distant galaxy is an old familiar story.

There is something fascinating about space flight. It begins with simple principles that keep the planets, satellites, and intruding meteors in their paths, principles that have changed the course of history. It tells of sailing ships on beams of sunlight that reach out toward the Sun or far beyond the comets. Of spaceships that move in great ellipses to Titan, the ice-swept giant moon of Saturn. It creates worlds of weightlessness and simulated weight and floating factories that fashion space ore into islands in the sky that rival Gulliver's.

Surprising as it may seem, it all can come to some good. To help solve Earthly problems gives it some purpose, and that is the purpose of this book.

CHAPTER ONE

A Trainful of Cobwebs

Uraniborg

There used to be an old ferry making the half-hour run to a small island called Hven from the fishing village where I live, three quarters of the way from Copenhagen to Elsinore. To reach the island now you have to sail clear across the Øresund to mainland Sweden and there take another ferry back to Hven. But it is well worth the trip. You can walk all around the island in a long afternoon, get some stunning views of Denmark to the west and Sweden to the east, and go swimming in sparkling clear water.

You will see no cars. Only bicycles, dirt roads, some houses, rolling meadows and forests, and a few people—mostly visitors who need a quiet day off from the bustle of Copenhagen and Malmö. Through my study window, at this moment, Hven floats in the morning mist like Jonathan Swift's Laputa, an emerald world bathed in sunlight, lying more in the clouds than on the sea. On Laputa, Gulliver met stargazers whose "discoveries" of the two moons of Mars cause present-day astronomers to wonder, just a bit, how Swift could have known.

If you walk to the middle of Hven you will encounter a ruin, Uraniborg, named after Urania, the muse of astronomy. And on the other side of the road you will learn its story in the one-room museum. For here, almost in the shadow of Hamlet's Kronborg, Tycho de Brahe built the greatest observatory the world had ever seen.

To some people, Tycho the Dane must have appeared quite mad. His nose was of pure silver and gold, his real one given up in a duel. But mad or not, his work marked the turning point of civilization, the transition from the ancient to the modern world.

Hypotheses and theories are always easy to set forth from a writing desk or an armchair. But meticulously made observations are often quite another story. Uraniborg was constructed around delicately pivoted, superbly calibrated long tubes through which the light of the stars and planets passed. Alas, the telescope was still to be invented, for this all predated Lippershey and Galileo by a full quarter of a century.

Some people place the real birthdate of modern science at 1543, the year that Nicolaus Copernicus' monumental work, *De Revolutionibus Orbium Celestium,* or *Concerning the Revolution of the Celestial Spheres,* was published as he lay on his deathbed. Copernicus dedicated his work to the Pope, but that did little to stem the reaction to his thesis, the heresy of the Age—that humankind's habitat, *terra firma,* was not indeed the center of the Universe, around which showed the nightly pageant of the stars, and the circling Moon, Sun, and wandering planets. Copernicus made a simple geometric transformation to make celestial wanderings easier to understand. He proposed that the Universe have a new center: "In the middle of all dwells the Sun. Who indeed in this most beautiful temple would place the torch in any other or better place than one whence it can illuminate the whole at the same time? . . . We find, therefore, under this orderly arrangement, a wonderful symmetry in the Universe, and a definite relation of harmony in the motion and magnitude of the orbs, of a kind it is not possible to obtain in any other way."

Thus, in his model all the planets, Earth included, move about the Sun in circles, each one larger than the one before. Perfect circles? Even Copernicus admitted the fallacy of this conception. His model was what scientists would call the first order of approximation. The idea of transforming the origin was alone the great leap forward beyond the Earth-centered model of Hipparchus from about 130 B.C. Expounded and recorded by Ptolemy of Alexandria three centuries later, the Greek model continued to hold the field through the Middle Ages.

The heliocentric model of Copernicus originated in fact with Aristarchus of Samos in the third century before Christ and was described a short time later in the work *The Sand-Reckoner* by Archimedes. No one is quite sure whether or not Copernicus was aware of Aristarchus' model. In Bertrand Russell's *A History of Western Philosophy* one reads: "Copernicus came to know something, though not much, of the almost forgotten hypothesis of Aristarchus, and was encouraged by finding ancient authority for his innovation." Later in the book, Russell writes that Copernicus "does not seem to have known of Aristarchus' heliocentric theory, but there is nothing in his speculations that could not have occurred to a Greek astronomer."

Like the Greek Pythagoreans, Copernicus takes it as axiomatic that all celestial motions must be circular and uniform, mainly for aesthetic reasons. However, to make his model approach reality he was forced to admit epicycles, circles which move about other circles.

What is reality? The nightly observations of the ancient Greeks. The shafts of light down Tycho's tubes. The falling apple. The rolling wheel. Wise men may argue all day about theories, but only fools will debate the facts.

To settle once and for all the conflict between the two theories—heliocentric and geocentric—the world awaited a clearer statement of the facts. The man to provide the scientific data was Tycho de Brahe. And the place was Uraniborg.

Not that Tycho believed in Copernicus. Quite the opposite. Tycho's armchair theory still held the Earth (and Man) to be at the center of all things. But Tycho's model had an original twist. To the Copernicans he conceded that the planets revolve about the Sun. There he stopped. He still wanted the whole bundle to move about the Earth.

Then why was Tycho such a key figure for modern science? Because, without really wanting to, he provided the very data which shattered all geocentric theories–his own included! It was the scientifically gathered data which marked the turning point, and all who argued against it were indeed fools.

World models

So began the avalanche which destroyed one myth after another. Tycho's disciple, Johannes Kepler, refined the Copernican circles to the next order of accuracy, perfect ellipses, or as near perfect as the measurements of those days admitted. By fitting ellipses to Tycho's data the difficulties with the heliocentric model were resolved and the Ptolemaic system, with all its theological ramifications, was buried in one blow. Separately in Italy Galileo developed a set of principles to describe moving objects on Earth, based upon his own experiments; this obliterated the myths in the world close to home. It remained for Isaac Newton to unite the terrestrial world of Galileo with the celestial world of Kepler into a great synthesis. The dichotomy of the ancients–two worlds, one on Earth, the other in the sky–gave way to one universal system. By the year 1700 the point of no return had been reached and what we now regard as the Modern Age had firmly taken hold.

The pace of observational astronomy picked up. The

appearance of Tycho's Star in 1572, which lacked a daily parallax and therefore must be more distant than the Moon, already began to make people suspicious of the teachings of Aristotle who maintained that change could occur only in the sublunary sphere. When Galileo pointed his new telescope at Jupiter he saw four bright satellites moving about the great central body—a Copernican world in miniature. The Galilean telescope led the way to the giant reflectors. From there to the spectroscopes, the interferometers, electronic image intensifiers, radio and radar telescopes, laser probes, and finally space platforms.

And it was inevitable that the idea of a fixed and absolute reference system centered in the Sun (or anywhere in space for that matter) joined the growing circle of myths. This time, however, the myths were Newtonian rather than Aristotelian or Ptolemaic. The answers to ageless questions: Where does space end? When does time begin? remained more elusive than ever.

Time itself, "as constant as the stars" in Newton's day, yielded to crucial experiments which appeared to show that it depended on the state of motion of one reference frame as observed from another. And those most fundamental of concepts—length and mass along with time—which are the essence of the only Universe which mortals know, appeared to lose all measure of objectivity, the *sine qua non* of science. They seemed to dwell more in the eye and touch of the beholder.

Newton's world system was likened to a great mechanical clock. But a clock is merely an analog, a model portraying reality. The Universe was reality itself, the ultimate perfection. Knowledge of it could be approached, but never really attained. Mere humans could peck away at its hidden secrets, pick up a few pebbles. The vast beach lay beyond. By the start of the nineteenth century this mechanical world view was so drilled into the scientific establishment that the radical departures from it demanded by new data—Einstein's Relativity Theory and Bohr–Heisenberg–Planck's Quantum Theory—became the new heresy.

The heresy held, and now it forms the accepted platform of modern science. Reality is not something apart from us, but is shaped by fundamental limitations in our capability of making measurements. Travel to remote worlds is limited not by their incredible distances, but by our ability to alter time and space in our means of travel. If four hundred years of science teaches us anything it is this: the "impossible" may not even be improbable.

I remember sitting in the Hayden Planetarium in New York shortly after World War II ended with the dropping of two atomic bombs over Japan. The lecturer tried to explain to his audience how the planets move around the Sun: "Just like electrons around the nucleus of an atom." Apparently the Rutherford–Bohr Solar System model of the atom was even more well known to the audience than the real Solar System. Rutherford and Bohr would have been the first to admit that their models were merely convenient ways to portray events in everyday terms that humans could understand. They soon gave way to a more accurate, but less accessible, representation of reality in the world of microphysics—hazy electron clouds ringing the central nucleus. The density of the clouds in space gave a measure of the probability density, the likelihood of finding an electron at a particular distance from the nucleus.

What model can we propose to impress upon us how immense are the distances to remote worlds?

Suppose we wished to build a planetarium of our own to the scale at which the Earth were the size of a walnut, say 2.5 centimeters (one inch) in diameter. How large would the Sun be, and how far away from the Earth in our model? It would turn out to be a globe over 2.8 meters (9.1 feet) in diameter, and 297 meters (970 feet) from our walnut-size Earth, measured center to center.

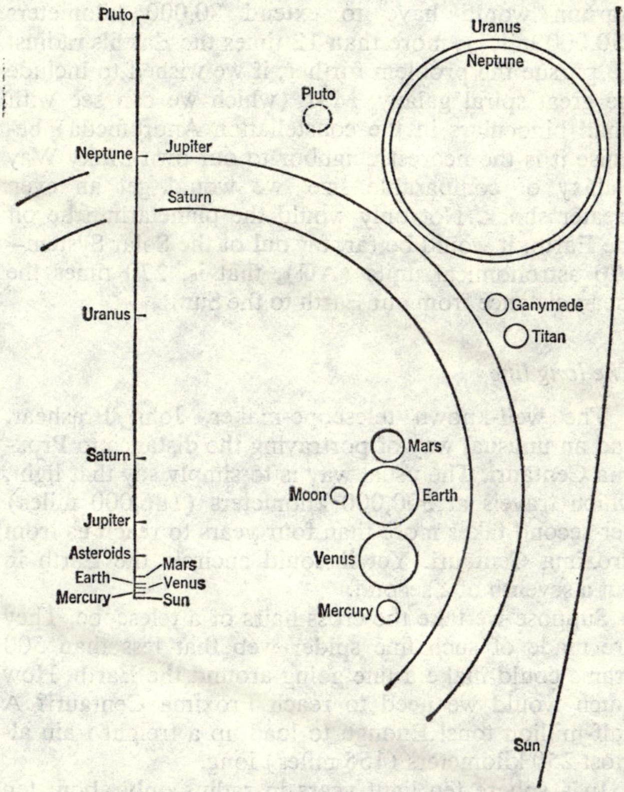

FIG. 1. The relative distances of the planets and their relative sizes. The average location of the asteroids is shown; actually they are spread between the Martian and Jovian orbits and sometimes lie beyond them. Pluto's diameter is not accurately known. Ganymede and Titan are the largest satellites of Jupiter and Saturn, respectively.

The planetarium would have to be 22.8 kilometers (14.2 miles) in diameter to include the whole Solar System up to the orbit of Pluto. But to reach Proxima Centauri, the nearest star to the Solar System, the planetarium would have to extend 80,000 kilometers (50,000 miles), more than 12 times the Earth's radius! To pursue the problem further, if we wished to include the great spiral galaxy, M31 (which we can see with small binoculars in the constellation Andromeda) because it is the nearest neighbor to our own Milky Way Galaxy of comparable size, we would get an even greater shock. Not only would the planetarium be off the Earth, it would be far, far out of the Solar System—270 astronomical units (AU), that is, 270 times the mean distance from our Earth to the Sun.

The long line

The well-known telescope-maker, John Brashear, had an unusual way of portraying the distance to Proxima Centauri. The usual way is to simply say that light, which travels at 300,000 kilometers (186,000 miles) per second takes more than four years to reach us from Proxima Centauri. Yet it could encircle the Earth in but a seventh of a second.

Suppose we take the cross hairs of a telescope. They are made of such fine spider web that less than 500 grams could make a line going around the Earth. How much would we need to reach Proxima Centauri? A half-million tons! Enough to load up a freight train almost 250 kilometers (155 miles) long.

In a sphere ten light years in radius only about ten stars are counted. In a radius of 100 light years, 10,000 stars are included. To the remotest stars of our galactic system, we must extend across 80,000 light years encompassing a hundred billion stars.

Such large numbers that are used to describe our cosmic neighborhood have a way of discouraging us and make us feel forever tied to this remote corner of space. Yet the future of space travel may hold surprises, ways to some day overcome what today seems "impossible."

Long ago Karl Compton, a former president of MIT, showed by a rather unique exercise how a phenomenon of everyday experience can involve scales of magnitude which dwarf even the dimensions of the cosmos. In the Smithsonian Institution report for 1937 he wrote:

> What is the largest number that has any physical significance? This is impossible to answer, being largely a matter of definition. But one common answer to this is 10^{110}, or one followed by 110 ciphers. This is about the number of electrons (the smallest things known) which would be required to fill up the Universe to the greatest distances discovered by astronomy, if the electrons could be imagined to be closely packed side by side to fill up this whole space. Yet this number, large as it is, is very small indeed compared with the aggregate factor by which the energy of a voice striking a telephone transmitter in San Francisco is amplified by electronic tubes in the process of a long-distance telephone conversation to London. This amplification factor is about 10^{256}, or unity followed by 256 ciphers. If the Universe were multiplied in size by the number of times it is larger than an electron, it could still not hold as many electrons as the number of this telephone amplification factor.

In the 1930's the Universe was thought to be about three billion light years in radius. At the present time 12 billion is the figure often given. But even if the radius were ten times larger than in Compton's time, then the number of electrons squeezed into it would be 10^{113} and into the super-Universe, as defined by him, 10^{226} ordinary electrons would be squeezed. This number is still *far smaller* than the number of times you would have to increase your voice level in San Francisco to be heard in London without benefit of electronics.

What were remote dreams in the 1930's have become reality two, three, and four decades later. Manned space travel to the Moon. Probes to most of the planets of the Solar System. Soon we will be moving far outward, to the domain of the comets, with sailing ships propelled by solar radiation as described in Chapter Five. And some day the stars beyond and their possible planetary systems will be within reach. In recent years we have found evidence of organic molecules in space, providing that needed hint of the possibility of life. How likely is it that we will find planetary systems not unlike our own scattered through space? I attempted to discuss this question and the related one: Why explore space? In the introduction to a book in space flight which was published on the very day that Alan Shepard made his first suborbital flight, 5 May 1961,* I wrote:

> Of all of life's experiences, one of the most moving is to stand in an open field on a clear crisp night, engulfed in a spectacle so staggering that it almost tears us away from our Earthly surroundings. We are plunged into the depths of a mystery that surely baffled even the first reasoning creature that lifted its head toward the sky. By centuries of observation and study, mankind has advanced a long way from the inverted-bowl concept of the ancients. We have weighed the stars, computed their distances, determined their motions, studied their compositions, measured their temperatures, and speculated on their life histories. With all this knowledge behind us, the mystery of the Universe still is beyond our comprehension. The compelling question: Where do we and

* Arthur I. Berman, *The Physical Principles of Astronautics* (New York: John Wiley & Sons, 1961).

our world fit into this colossal structure? remains the great riddle of this age as of the ages past.

The central cultural event of the twentieth century has been the impact on human life of pure and applied science, from quantum and relativity physics to solar-powered satellites. Every facet of thinking and living has been affected: social, political, philosophical, economic, and even artistic. The science of astronomy, too, has leaped ahead in strides characteristic of the age. The great telescopic cameras (notably the Hale and Schmidt instruments on Mount Palomar), for sheer immensity and delicate precision rivaled only by the high-energy nuclear accelerators now being built, the radio telescopes that have sprung up over the world, and the instrument-carrying balloons, rockets, and satellites have revealed a celestial structure heretofore completely invisible.

Where have the advances of modern astronomy brought us in our attempt to understand man's position in the Universe? The monumental pronouncements of Copernicus, Kepler, and Newton revolutionized the thinking of the seventeenth-century educated person. Now, his twentieth-century counterpart is forced to undergo a new upheaval in thought, supported by evidence which, while more subtle than that provided by the giants of post-Renaissance astronomy, is even more dramatic in its significance.

Harlow Shapley distinguishes four great adjustments in the history of humankind's quest for orientation in the Universe. The first adjustment involved the early realization that the center of the visible Universe was not the point on the ground at which a person stood. Thus the geocentric concept, associated with Ptolemy but actually proposed centuries earlier by Hipparchus, replaced the anthropocentric. The second adjustment, the heliocentric theory of Copernicus, a revival of the idea of Aristarchus, was accepted only with reluctance, as it humbled humankind's position as did no other concept before. Then, with the advance of astronomy, came the third adjustment: a gradual realization that the Sun is but an ordinary star in a spiral arm of a rather commonplace galaxy, one of billions of galaxies.

The abasement of the Earth and Sun and the elevation of the galaxies do not constitute the final chapter in this sequence of reorientations. The fourth adjustment lies in the realm of biology, aided by the findings of astronomy and chemistry. Shapley gathers together such cosmical facts as the expansion of the Universe, the laws of motion of celestial bodies, and tidal forces and arrives at an estimate, conservative in the extreme, that a star has circulating about it a planet meeting all the physical requirements for life: one star out of a million million. To anyone unfamiliar with the major facts of modern astronomy, this might appear to doom the likelihood of finding life beyond the limits of our Solar System. But we then couple this low probability with the totality of stars in the presently known Universe, a number so large that it defies human comprehension: 10^{20}. The resulting inference, that there are no less than a hundred million planetary systems capable of supporting life as we know it, casts a shadow of doubt on the proposition that the Earth, the Solar System, or even our Galaxy somehow is unique.† Bringing to bear our present knowledge of complex molecules and the structure of simple organisms, we are drawn inescapably toward Shapley's conclusion: "We are not alone."

† In 1961 a "hundred million planetary systems capable of supporting life as we know it" was a startling figure. Studies made in 1978 of the angular momenta of nearby stars have now yielded an estimate of the number of such planetary systems in our Galaxy alone of *ten billion.*

> It is not certain that the sum total of human happiness will be increased by interplanetary, interstellar, or intergalactic explorations. Perhaps there will be found no rare gems, no new energy sources, no areas for colonization, not even a measure of military security. Yet, if we slowly begin to remove the bonds that tie us down to this remote speck in a corner of one of the countless galaxies scattered through the Universe, it may be that we will gain understanding of the world that surrounds us, of our place in it, and of ourselves.

Why go to the stars? Alfred North Whitehead summed it up this way: "Without adventure civilization is in full decay."

CHAPTER TWO

Stop the World, I Want to Get Off!

Moon fare

How much does it cost to go to the Moon? In the mid-seventies an Apollo flight cost about a third of a billion dollars. On the other hand, some people have pegged the price of Apollo 11 alone at 24 billion dollars. It depends on the accounting method that is used. If you computed the cost of the first atomic bomb, the one that was tested in the New Mexico desert at Alamagordo in July 1945, and included all the research and development work, it would have come to two billion dollars, which was quite a lot of money in those days (and still is!). On that date it was the only such bomb in existence.

But how much *should* it really cost to reach the Moon routinely, expressed in total energy costs at current prices? To leave the Earth completely means that we have to push our rocket against a force of attraction that is strong close to the Earth and gets weaker and weaker the farther out we go. It never really drops to zero even at very great distances, but for all practical purposes we can assume that it is zero.

Isaac Newton, with the knowledge of Kepler's laws of planetary motion behind him, described the force of gravity in a very simple way. He said, first, that it is directed toward the center of the Earth, and second, that if you are on or above the Earth and you double the distance out from the center, the pull of gravity drops by a factor of four. If you double it again it drops by a factor of 16, or as a general rule, it weakens in proportion to the *square* of the distance to the Earth's center.

When the phrase "pull of gravity" is used here it means the force of the whole Earth on a fixed object, say a one kilogram mass either in motion or at rest, its distance at each instant referred to the Earth's center. Newton noted that each tiny element of matter comprising the Earth has its *own* attraction to that kilogram mass, unaffected by its neighboring elements (i.e., there is no *shielding* effect of one element upon another). The force between the element and the kilogram is directed along the line between them and depends also on the distance between them. Curiously enough, if you add up the attractive effect of all the Earth's elements upon that single kilogram, it is as if Earth's entire mass were concentrated at a tiny volume at its center. This conclusion isn't at all obvious and is subject to a slight error. The error is not found in the analysis but in the peculiar way that mass is distributed within the Earth, which really isn't spherically symmetric after all.

It is fairly easy to demonstrate that if each particle of mass which makes up a thin spherical *shell* attracts the kilogram mass (also considered to be small compared with the dimensions of the shell) according to the Newtonian inverse-square law, then the whole shell attracts the kilogram as if the shell were a single particle at its center. Our only requirement is that the elements of mass making up the shell are uniformly distributed. We know that the *density* (the mass of a volume element divided by the size of the volume element) of the Earth is not uniform throughout but gets greater as we go to the center. But that would make no difference. So long as the Earth could be assumed, as a first approximation, to be made up of many thin shells, each of which is of uniform density, we may treat our planet as a point mass placed at its center as far as its gravitational effect, and therefore the motion of its satellites, is concerned.

It is interesting to observe that if our standard kilogram mass were anywhere *inside* a thin spherical shell, no matter how massive, its total gravitational effect on the kilogram–that is, the total attraction on the kilogram of all the small mass elements comprising the shell–would be *zero!* This would mean that if we were deep inside the Earth, only that mass of the Earth below us, i.e., contained in a sphere of radius equal to our distance to the center, would play any role in producing a net gravitational attraction.

The departure from the simple conclusion that the Earth attracts a satellite as a point mass is caused by two factors: First, the Earth's density depends not only upon radial distance from the Earth's center, which does not affect the satellite, but also upon latitude and longitude, which does. Second, the shape of the Earth is not at all a perfect sphere; it is flattened at the poles and bulges at the equator, caused, of course, by the Earth's daily rotation.

Even if the Earth were able to maintain its rigidity as a perfect sphere despite its rotation, we still would have a problem. If we attempted to measure the force of gravity by, say, placing an object on a bathroom scale, we would get the illusion that the force of gravity is weaker than it would have been if the force were based on the gravitational attraction of the Earth's bulk alone, measured on a freely falling object–a falling apple or an orbiting satellite. Imagine the following "thought experiment": the Earth rotates with a period of 1½ hours instead of 24 hours. The weight of objects "resting" on the Earth's equator will have dropped to zero; they will be *weightless* and truly in circular orbit, ignoring the drag force effect of the air.

But rather than unnecessarily complicate our argument at this point, let us assume our kilogram mass to be placed upon an idealized Earth–one which is perfectly *spherically symmetric,* that is, appearing the same in all directions as far as its distribution of density is concerned. We have indicated that such an Earth gives rise to a measurable gravitational "force field" in space which obeys the "inverse-square" law of gravitational attraction as referred to the Earth's center.

If we measure the attractive force on our kilogram mass when it lies on the Earth's surface, we can call it g_0. If we move the kilogram twice as far out–to an altitude equal to the Earth's radius–the attractive force

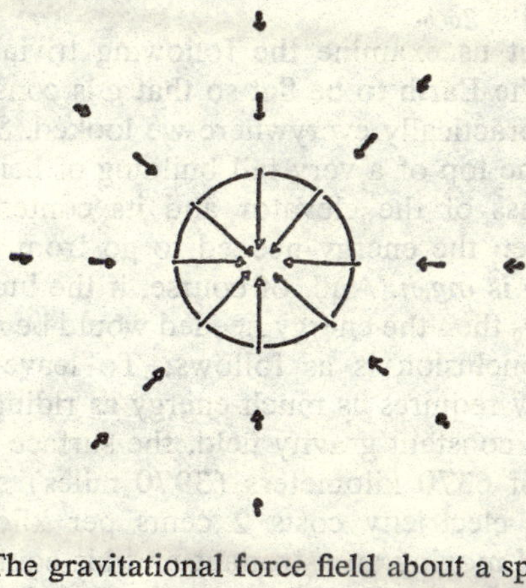

FIG. 2. The gravitational force field about a spherically symmetric mass is directed toward its center–a one-kilogram mass placed anywhere in space will be pushed in the direction of the arrow and the size of the push depends upon the length of the arrow. This size obeys the *inverse-square law* of Newton, that is, it varies as $\frac{1}{r \times r}$ or $\frac{1}{r^2}$, where r is the distance to the center.

now will be only $\frac{1}{4}g_0$.* As we see, when we return back to the surface so that $r = r_0$ then $g = g_0$ as it should.

It is a fairly straightforward problem in integral calculus to find how much *energy* is needed to push the 1 kilogram object to some very distant place far from the Earth, starting from the Earth's surface. (Or, what is equivalent, the work done by the attractive gravitational field, if it should pull the object in from a great distance r down to the surface at r_0.) We have to add up the little "elements of work" as we push through each small distance, first against the strong pull of gravity close to the Earth, and later against weaker and weaker pulls as we get it further and further out. The result of the calculation of the energy needed to get it all the way out (we have assumed that r is so large that gravity there is simply too weak to require our pushing measurably against it) is g_0r_0.

Now let us examine the following trivial problem: Imagine the Earth to be flat so that g is constant at the value g_0 practically everywhere we looked. Suppose we went to the top of a very tall building of height h. If m is the mass of the elevator and its contents in kilograms, then the energy needed to go from the bottom to the top is mg_0h. And, of course, if the building were as tall as r_0 then the energy needed would be mg_0r_0.

Our conclusion is as follows: To leave the Earth completely requires as much energy as riding in an elevator in a constant gravity field, the surface value g_0, a distance of 6370 kilometers (3970 miles) straight up. Base-load electricity costs 2 cents per kilowatt-hour; the cost of moving an astronaut and his equipment, say 1000 pounds, or 454 kilograms is now found as fol-

* At *any* radial distance r, we can write the general expression for the inverse-square gravitational attraction on our standard kilogram object:

$$g = g_0 \frac{r_0^2}{r^2},$$

where r_0 is the Earth's radius.

lows: 3970 miles is 2.1×10^7 feet so 2.1×10^{10} foot-pounds of energy are needed. Now one kilowatt is 550 foot-pounds per second, so one kilowatt-hour is 550 foot-pounds/second $\times$ 3600 second/hour $\times$ 1 hour = 2.0×10^6 foot-pounds. Thus we require

$$\frac{2.1 \times 10^{10}}{2.0 \times 10^{6}} = 10{,}500 \text{ kilowatt-hours}$$

at a total cost of $210 or 21 cents a pound! This is a far cry from the cost of Apollo, but it is a goal to strive for.

Escape speed

The practical means for our astronaut and his equipment to leave our spherical Earth is far more complicated than a trip in an elevator. He leaves in a rocket, which amounts to exploding some propellant inside a container open at one end and moving the container by the energy of the explosion. We want this explosion to take place at as low an altitude as practicable, for the last thing we want is to lift unexploded propellant to high altitudes unnecessarily, to be carried aloft by the portion that has been exploded. There is simply not a propellant yet known which has enough chemical energy to lift its own weight into orbit, let alone escape Earth's gravity completely. To orbit a rocket—the empty container together with its payload of astronaut and equipment—requires ten times its weight of propellant, exploded close to the ground.

This, however, would not be the case if one were to build a nuclear rocket. Here the energy released is far more than is necessary to carry the bulk of nuclear fuel toward the stars, but the gap between theoretical expectation and practical hardware is enormous. To build a reactor, surround it by tons of shielding, and convert it into a working rocket is a tall order. For one thing, the final space vehicle would be perhaps a hundred times more massive than the fuel alone, the reverse of the situation we have with the chemical rocket. For another, we are not sure whether it is better to drive the fuel itself through the rocket nozzle, or expel some other material instead (e.g., hydrogen gas) that first absorbs energy from the intensely hot reactor. The environmental hazards that would be present, especially if lift-off doesn't occur properly, are of course additional problems.

As in the case of hurling a stone, we require as much initial speed as possible before release. How much speed is actually needed to get it as far out as possible from Earth's gravity? We know that its initial *kinetic* energy—energy of motion—depends on the *square* of the speed and is written

$$\tfrac{1}{2}mv_e^2.$$

The energy needed to push the object, of mass m, all the way out against the pull of gravity toward the Earth, we have stated to be

$$mg_0r_0.$$

The subscript e means that the speed, v_e, is large enough close to ground level to enable the object, a rocket for example, to completely *escape* the gravity field of the Earth. By definition, the two energies will equal each other, and so

$$v_e=\sqrt{2g_0r_0}.$$

(We must bear in mind throughout this discussion that air resistance is not to be reckoned with at this point; it would needlessly complicate the problem.) Thus a stone thrown at the escape speed from the Earth's surface will never fall down.

Two things stand out at once: First, whether the rocket or the stone is very massive or not is irrelevant. The required escape speed is the same. Second, nowhere in the analysis did we have to specify the direction of launch. Our intuition might tell us that we must launch straight up, but in fact we can launch horizontally, or at any angle between the horizontal and vertical. The projectile still will never return to Earth.

We can even *reverse* the dynamics. That is, if we released an object from a very great distance from the Earth, being very careful not to give it any motion with respect to the Earth, the object will be pulled down by the then infinitesimal gravity field of the Earth, hardly

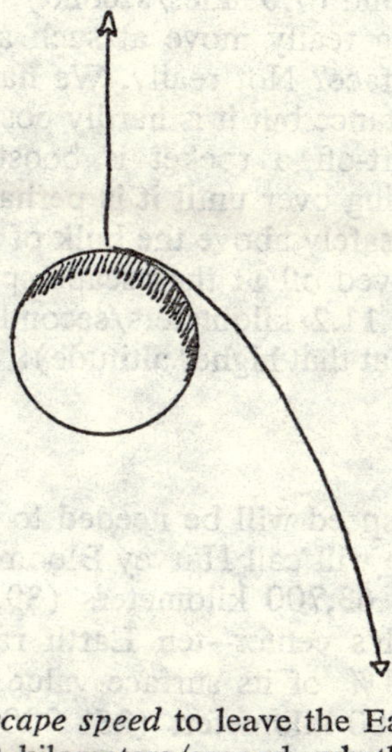

FIG. 3. The *escape speed* to leave the Earth at low altitudes is 11.2 kilometers/second, whether we launch the object horizontally or vertically. There are practical reasons for vertical lift-off of a rocket and then bending over into horizontal flight: (1) It cuts through the dense lower atmosphere in as short a path as possible. (2) By bending over toward the *east* full advantage is taken of the speed of the Earth's rotation; the rocket's speed in space is the sum of its speed relative to the launchpad and the pad's speed in space.

moving at all at first. Then it would fall faster and faster as the attractive force field gets stronger and stronger. At last it will crash into the Earth and the speed of impact will be v_e—the total energy imparted to it by the gravity field from "infinity" all the way to the surface appears as kinetic energy at impact.

What is the actual value of v_e at or near the Earth's surface? We know the strength of Earth's gravity at or near the surface—9.8 newtons/kilogram, which may also be written 9.8 meters/second^2.† And the Earth's radius is 6.37×10^6 meters, so v_e turns out to be 11.2 kilometers/second (7.0 miles/second).

Can anything really move at such a great speed at the Earth's surface? Not really. We have continued to ignore air resistance but it is hardly possible to do so in practice. At lift-off a rocket is boosted straight up, gradually bending over until it is perhaps one hundred kilometers up, safely above the bulk of the atmosphere. There it is shoved off at the escape speed (which is a trifle less than 11.2 kilometers/second because of the weaker gravity at that higher altitude).

Orbital speed

The escape speed will be needed to carry our astronaut, whom we will call Harvey Bloom,‡ to the Moon, or beyond. At 63,700 kilometers (39,700 miles) out from the Earth's center—ten Earth radii—the pull of gravity is but 1% of its surface value. At the Moon's distance, 384,400 kilometers (238,900 miles), or 60.3 Earth radii, the Earth's field is less than 0.03% of its

† We have the choice of measuring the force directly as a pull, with a spring balance for example, or we can measure its effect in causing an unheld object to fall freely and accelerate. The latter is better, for the spring balance would respond to the Earth's rotation which has nothing to do with our problem.

‡ The hero of Allan Sherman's song: Shine on, shine on Harvey Bloom/Up in the sky/You have been in orbit/Since January, February, June and July.

surface strength. While the Moon's gravity is only 16.5% of the Earth's surface gravity it still is 600 times stronger at its own surface than the Earth's gravity

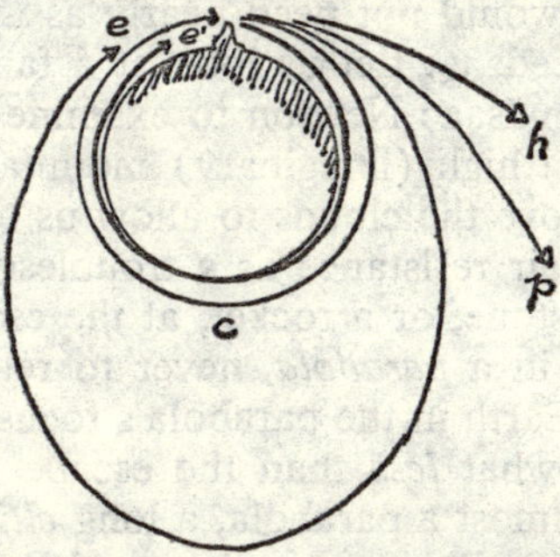

FIG. 4. If an object were cast from the Earth horizontally at the escape speed v_e it will move in a parabolic arc p as shown. At slightly less than the escape speed it will move in a closed orbit (a long ellipse e) and repeat its motion periodically. The period will be large if the ellipse is long. The greater the "cast-off" or "burnout" speed, in the case of a rocket, the greater will be the length and period of the ellipse. If the object were launched horizontally at a much lower speed —the "circular orbital speed", v_c—the ellipse will now be a circle and the period of the orbit will be the circumference, $2\pi r_0$, divided by the speed, v_c. If the object were launched at less than v_c, a shorter ellipse, shown as e', would be formed. If this ellipse strikes the Earth the path will be suborbital. If the object were cast off at a speed greater than the escape speed the path would be *hyperbolic,* denoted by h. Like the parabolic case, the object in a hyperbola would never return to Earth. In the parabola the object approaches an infinite distance with infinitesimal speed; in the hyperbola it approaches an infinite distance with finite speed. The analysis is simplified in that perturbing effects, such as air resistance, are ignored. (Adapted from an original drawing by Isaac Newton.)

there. Because of this, a rocket from the Earth to the Moon must carry along reserve propellant to land Harvey on the Moon in a soft landing.

If we merely wanted to orbit Harvey Bloom *about* the Earth we would not need nearly as much speed as the escape speed, v_e. Let us look at a famous drawing borrowed from Isaac Newton to examine the problem. We stand on a high (imaginary) mountain which juts far enough above the clouds to allow us to forget completely about air resistance as a troublesome factor. If we cast off a stone, or a rocket, at the escape speed, it will move off in a *parabola,* never to return, with the center of the Earth in the parabola's focus. If we cast it off with somewhat *less* than the escape speed, it will move out in almost a parabola, a long *ellipse.* This is a closed curve; it will return to its starting point and the motion will repeat itself along the same path. The Earth's center will be in one of the two foci of the ellipse.

If we should instead cast off the stone at a lesser launch speed it will move in an ellipse, but a shorter one than before. With even less launch speed, the ellipse shrinks in size still further. At last we come to a critical launch speed which will carry the spacecraft into a circle about the Earth, maintaining this speed at all times. What is the value of this speed, which we will call the circular speed, v_c, and how does it compare with v_e?

To find the circular speed we ought to know something about the relationship between force and acceleration. Aristotle taught that the only way an object can continue moving is for something to keep pushing it. In a way this made some sense, for everyone knows that if you slide a block across the floor it will come to rest unless you keep it moving. You have to keep your foot on the gas pedal of a car just to keep it moving steadily at 100 kilometers/hour. Aristotle was responsible for much of the erroneous thinking about physics that persisted throughout the Middle Ages. What he did not consider was the force which was already pressing on the block and the car, the force of friction pushing back against the forward motion. This force would decelerate both the block and the car to bring them finally to rest, were it not for the counterforce–our hand on the block or exploding gasoline against the pistons in the engine cylinders, transmitted to the wheels–which

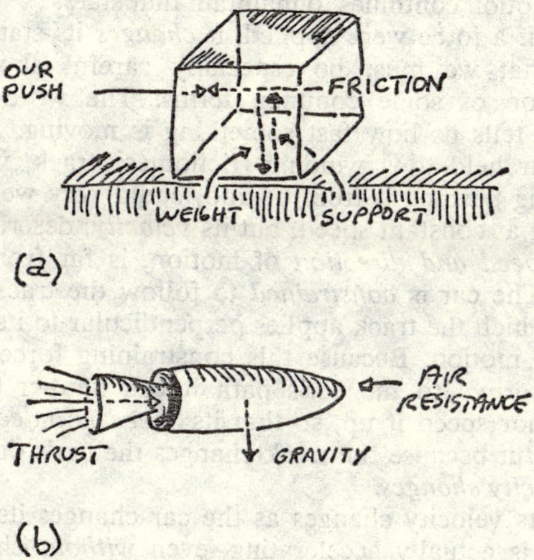

FIG. 5. (a) Just to keep a block sliding along at constant speed means that you have to keep pushing it to balance out the force of friction. In the vertical direction the object's weight and the resisting force of its support are also in balance.

(b) A rocket's acceleration will, at any moment, determine its future motion based upon its past motion. The acceleration itself is determined by the *reaction force* of the emitted propellant as well as the resisting force of air and the combined gravity forces which may be present.

cancels the force of friction and keeps them steadily moving.

If you imagined a world without friction, like airless space, and a rocket with its propellant depleted cruising in a region where gravity is so weak it cannot be detected, then no force whatsoever is needed to keep it moving along. It will maintain its speed as well as the direction of its motion. This is the principle of *inertia:* In the absence of a net applied force, an object at rest or in motion continues to maintain that state.

But if a force were applied, it *changes* its state. It is here that we must be especially careful about the definition of some common terms. The word *speed* merely tells us how fast something is moving. A railway car held to a winding frictionless track, without anything pushing it from behind, could very well keep moving at constant speed, but its *velocity,* described by *both speed and direction* of motion, is far from constant. The car is *constrained* to follow the track, by a force which the track applies perpendicular to its direction of motion. Because this constraining force is exactly *normal* to the car's path it can neither slow it down nor speed it up, so that its *speed is indeed constant.* But because the track changes the car's direction its *velocity changes.*

As its velocity changes as the car changes its direction it is actually accelerating—even *without* changing speed. The acceleration is always aligned with the constraining track force, that is, normal to the track, for it is the only force presumed present. The greater the curvature of the track the greater the magnitude of both the force and the acceleration, which we recall is the rate of change of velocity with time (obviously, the velocity change is great as the change in direction is great). Despite all these changes, the kinetic energy, based entirely upon its speed, *never changes.*

Although the magnitude of the acceleration is sensitive to the bending of the track, that is, the radius of

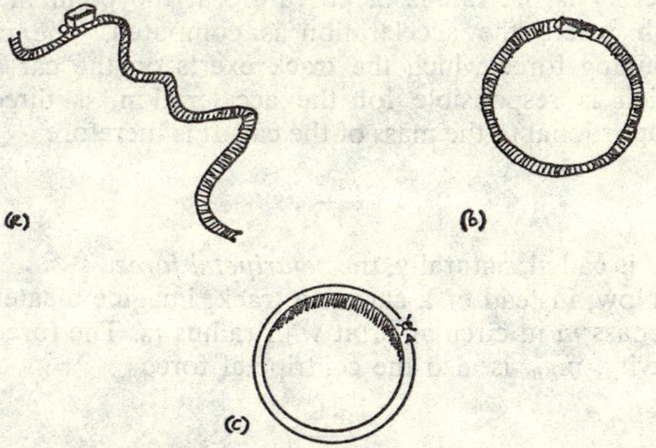

FIG. 6. (a) Even though this car may be on a winding track without friction and can therefore keep its speed constant (assuming nothing is pushing it or pulling it from in front or behind), it still is subjected to a strong force: the *constraint force.* This force always is *normal* to the direction of motion and therefore cannot speed it up or slow it down. It acts to keep it on the track and is therefore strongest when the curvature is greatest, for it is here that the departure from its natural straight-line motion is greatest.

(b) The train on a circular track is *continuously* accelerated toward the center in an amount $\frac{v \times v}{r}$, or $\frac{v^2}{r}$.

(c) A satellite in circular orbit is constrained to this path, given just the right launching condition, by gravity which is directed toward the Earth's center. Gravity acts here in the same manner as the constraining force of the track on the car in that it does not change the satellite's kinetic energy or speed.

curvature, it is even more sensitive to how fast it is moving. It has the name *centripetal,* or inwardly directed, acceleration. Its magnitude is simply

$$\frac{v^2}{r},$$

where r is the radius of curvature at the point in the path where the acceleration is computed. The constraining force which the track exerts on the car and which is responsible for the acceleration, is directly proportional to the mass of the car. It is therefore

$$\frac{mv^2}{r},$$

and is called, naturally, the *centripetal force.*

Now, instead of a car on a track, imagine a satellite of mass m in circular orbit with radius r_0. The force of gravity, mg_0, is also the centripetal force

$$\frac{mv_c^2}{r_0},$$

which holds it in the circle and prevents it from flying off in a straight line. It maintains the constant circular speed, which is

$$v_c = \sqrt{g_0 r_0},$$

or 7.9 kilometers/second (4.9 miles/second) for a low Earth orbit.

So, of course, it is easier to place something in circular orbit than it is to remove it entirely from the gravitational field of the Earth, for now we see that

$$v_e = \sqrt{2}\, v_c.$$

The length of time to complete the circular orbit is called the satellite's *period.* Since we know how fast it is going, v_c, and how far it must go in a circle, $2\pi r_0$, then the period is

$$\frac{2\pi r_0}{v_c}.$$

One would expect that if the circle were larger it would take longer to go around, i.e., the period would be greater; but how much greater? We may write, for any radius, r, that T, the period, is

$$T = \frac{2\pi r}{v},$$

where v is the speed. The speed depends on the gravitational field at that radial distance,

$$v = \sqrt{gr},$$

but g itself depends upon r; it weakens with the square of the distance to the circle's center—the inverse-square law.

This gives us a most important law for space mechanics, as well as dynamical astronomy, that the period at any radius is

$$T = \text{constant} \times r\sqrt{r}.$$

Another way to write $\sqrt{r}$ is $r^{1/2}$, so that $rr^{1/2}$ is really $r^{1\,1/2}$ or $r^{3/2}$. This relation between period and distance is the famous Kepler third law of celestial mechanics for the special case when the orbit of a planet or satellite is a circle rather than ellipse. When applied to ellipses, instead of "r" we use half the length of the ellipse; this is called the *semimajor axis* of the ellipse.

To reach the Moon in a long ellipse about the Earth, neglecting the Moon's gravity and other disturbing influences (called *perturbative effects*) using Kepler's third law and the semimajor axis of the ellipse (half the sum of the Moon's distance, as measured from the Earth's center, and the Earth's radius) takes *120 hours.*

If we launched straight up so that we just barely reached the Moon we could do it in 116 hours, but this would be a foolish strategy. As we soon will see, by launching toward the east the rocket can be aided by the Earth's rotation and use less propellant. The launch speed needed in the vertical case is 11.1 kilometer/second, a trifle under the escape speed (or *parabolic speed*) needed to take the rocket to infinity.

The horizontal launch, neglecting Earth's rotation,

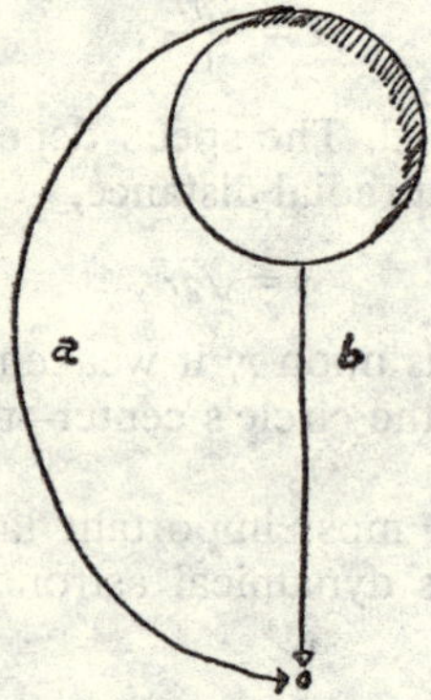

FIG. 7. Alternative orbits for reaching the Moon, neglecting the distorting effect of its gravitational field. (a) The half-ellipse, launched horizontally at the nearest point of the orbit to the Earth, called *perigee*. It takes 120 hours to reach the Moon, at *apogee;* this is found by extending Kepler's third law (of planetary motion about the Sun) to satellite motion, in this case rocket payload, about the Earth as the central body.

(b) A straight line, vertically up, part of a *degenerate* ellipse (or even a degenerate parabola or hyperbola). The time of flight is slightly less, 116 hours.

requires only 1.5 meter/second greater speed than that of the vertical launch. However, as the Earth's rotational speed is 0.5 kilometer/second at the equator, a launching to the east means that the rocket need carry sufficient propellant for a 11.1 − 0.5 = 10.6 kilometer/second launch. There is a further advantage to the horizontal launch as will be discussed in the next chapter.

Kepler's third law for a satellite in a low circular orbit gives a period of 1½ hours. This is why objects on our hypothetical Earth which rotated with a 1½-hour period could not rest on a scale—they would be in orbit, or weightless. We can also find, with this *3/2-power* law, at what height a satellite, placed in orbit above the Earth's equator, would appear to stay always at the same point in the sky, night and day. It must have a period of 24 hours; Kepler's third law gives us the radius of its orbit as 40,500 kilometers (25,000 miles).*

* This is found from:

$$\frac{24 \text{ hours}}{1.5 \text{ hours}} = \left(\frac{r}{r_0}\right)^{3/2}, \text{ or } \left(\frac{24}{1.5}\right)^{2/3} = \frac{r}{r_0}.$$

Here we see that we took the 2/3 power of each side of the equation to solve for the radius of the orbit in terms of the radius of the Earth.

CHAPTER THREE

Angular Momentum Conservation (With a Little Bit of Love) Makes the World Go 'Round

If you stand on a hilltop during a night so clear that the Milky Way appears as a great white belt speckled with millions of stars (though indeed you would need Galileo's telescope to see them separately), take a look to the constellation Cassiopeia, the "lady in the chair." She sits eternally facing the Great Dipper across Polaris, the Pole Star. Beyond Cassiopeia, an equal distance away from Polaris, shine the four stars that make up the Great Square of Pegasus. If you now draw an imaginary line between the closest star of Pegasus and that of Cassiopeia and bisect it, you will see close by a row of three stars at a 60° angle growing dimmer as you move away from the line. Look between the dimmest of the stars and the line–you will see what is surely the most monumental object in the heavens, apart from the Sun and Moon, that is visible to the unaided eye and certainly the most distant.

Here before you is an "island universe," the Great Galaxy of Andromeda, twin (or big brother?) to our home Galaxy, the Milky Way. Of course, to see it well you will need binoculars. You get the feeling at once that it is far off, far beyond the surrounding stars. And with a small telescope the details become clearer–a giant pinwheel inclined at an angle to us. A spectroscope, analyzing the *Doppler shift* of spectral lines, on one side toward the blue, on the other toward the red, would reveal that this strange, floating island of a hundred billion stars, 2 million light years away, is indeed *rotating.*

If you could probe beyond Andromeda with a large reflecting telescope into the realm of the galaxies, millions, billions of light years distant, not only would they be found to rotate about themselves, but also to revolve about one another in groups and supergroups.

Undisturbed rotation and revolution stays constant, from the minuscule particles in the inmost depths of the atom to the galactic supergroups. There are some things which are basic attributes of the Universe: length, time, mass, electric charge and force, magnetism, gravity, and energy. Angular momentum is another of these attributes.

What is angular momentum and the principle of angular momentum conservation? And why does it play such a fundamental role in physics, especially in space dynamics? To understand this we must backtrack to two concepts which are even more basic–linear momentum and its conservation, and the principle of inertia.

Idealizations in physics

Here again, in Figure 8, is the object introduced in Chapter Two moving along through space, straight as an arrow and at a constant speed, undisturbed by anything at all, not even gravity. Of course, this situation is only hypothetical, for gravity is everywhere. Even a thousand light years from the Sun, the feeble remnants of its gravitational field can, in principle, disturb, ever so slightly, the detailed path of a moving object. In principle perhaps, but in practice the disturbance is far

FIG. 8. An arrow, moving in a region of space where gravity, radiation, magnetic forces, and collisions by atoms, molecules, or meteoroids are negligible, will move along at constant speed and in a fixed direction.

beyond the range of our most sensitive instruments. So the remoteness from a source of gravity enables us to idealize motion as straight-line for objects which otherwise are undisturbed.

Purists still may quibble, for in the Einsteinian universe no line is truly straight. But physics can proceed only by describing idealizations which, in the real world, become merely approximations. How then could Galileo challenge the Aristotelians by stating that a heavy and light object will fall down at the same rate of speed? His critics missed the point entirely with their challenge: "Will a stone and a feather really fall down together?" The convincing thought experiment to satisfy the skeptics would have been to let two balls fall down side by side and then hook them together to form one object twice as heavy. Nothing fundamental would have changed which could have possibly influenced the mechanics.

An understanding of how a phenomenon proceeds ideally and how it is altered by real effects is the cornerstone of progress in theoretical science. Without this understanding we would have merely disorder and confusion. Sometimes we assume the Earth to be flat in order to help us solve various types of ballistic problems with an accuracy sufficient for many needs. Air resistance as a perturbing influence may be examined separately, both for missiles and feathers.

However, in those types of problems where the expected range of a thrown object is an appreciable fraction of the Earth's diameter, we must describe the gravity force field as pointing radially in toward the center of an Earth that is considered to be spherical. Yet even here we are making an approximation, for the gravity field is not spherically symmetric at all, as we know, but is distorted by the Earth's flattening at its poles and bulging at its equator. Yet it is not even "cylindrically symmetric," that is, the same in any direction about the polar axis, for the Earth's gravity field is actually very complex and asymmetric. Its actual mathematical description is revealed by the careful study of the motion of satellites under its control, and only after we have separately accounted for, and subtracted, the effects on the satellite's motion of such perturbing influences as air resistance, the Earth's magnetic field, solar radiation, and the gravity fields of the Moon, the Sun, and other celestial bodies.

Linear momentum

So when we describe the motion of our hypothetical object as undisturbed in a straight line, this is very much an idealization. Galileo, by his law of inertia, and Newton, in his first law of motion, each maintained that an object moves in a straight line precisely *because* it is undisturbed—nothing is pushing from the side. And more than that, it moves at constant speed—nothing pushes from behind or in front. Therefore we express this constancy of *both speed and direction* as the constancy of *velocity.*

Along with this idea of constant velocity, of undisturbed masses, we add another: If the *mass* of our moving object does not change—and why should it?—then the mass-velocity product, which we call *linear momentum,* is of course also conserved in the absence of an applied force. Conservation would also hold in the absence of a *net applied force*—a resisting force (e.g., friction) that cancels a push from behind has the same effect on linear momentum as no force at all.*

But if there *is* a net applied force, the rate of change of linear momentum per unit passage of time will follow it directly. To be specific, the *extent* of the change and *direction* of the change follow the magnitude and

* Net applied force is the force that remains when an impressed force on an object is partly canceled by an opposing force. Thus it represents the sum of the forces when the opposing force is regarded as negative.

direction of the force, respectively. This is the essence of Newton's famous second law of motion. We have seen in the last chapter how motion in a curved path could be caused by the constraining force of a track on an object, or gravity on a satellite in orbit.

Now we come to a subtle point. I asked above: Why *should* the mass change? It could, in fact, change in two ways: First of all, it could lose a part of itself, thus forming a system of two or more masses, and requiring us to study the dynamics of a multiparticle *system*. Sometimes we pay little attention to the mass that is lost. When a balloon loses air, what remains is still called a balloon. When a rocket expels propellant, what remains is still called a rocket. What really happens would be better understood if we referred to a "balloon-plus-air system" or a "rocket-plus-propellant system," where the total mass of each system (including all the molecules of air or rocket exhaust) remains constant as time goes on. Furthermore, the *total* momentum or velocity of the whole system would remain unchanged if there were no net force on the system. If there were a net force on the system, e.g., gravity or air resistance, then the *system as a whole* accelerates in response to this force.

Although Newton's laws of motion apply only to the dynamics of single particles in response to forces on them, we can derive principles from the laws which extend to systems of particles. We can separate the dynamical problem into two parts. One considers the interactive forces within the system, the force of one particle upon another; the other considers the force on the system, that is, the force external to the system, as applied to the *center of mass,* or average location of the masses which comprise the system at each instant. The outside force, such as gravity, is applied at the center of mass and causes the system to accelerate as if all the mass were located exactly there.

The second way in which a mass could change involves a concept which was as offbeat in the early part of this century as those of Galileo and Newton were considered in the seventeenth century. It is worthwhile to discuss this concept, even if briefly, for there is much experimental evidence to justify it, and it may well provide the impetus for planning voyages to those reaches of space, the distances of which are measured in light centuries!

A relatively small digression

There are several theories of relativity. The Newtonian theory of relativity states that if you have two *frames of reference,* say a train and a platform, moving at constant velocity relative to each other, it will be impossible to make any mechanical experiments on either, e.g., swinging pendulums, colliding balls, etc., which will determine which is really moving and which is standing still. It is not fair to merely look out of the window because you can never be completely sure whether or not the whole world is simply moving backwards and the train standing still. Reliance on intuition and the "pointer readings" of measuring instruments are two distinctly different things.

Albert Einstein went beyond mechanics into electromagnetic phenomena and stated that no physical experiment *at all* will provide even a clue as to which frame is "really" moving and which is "really" standing still. He was drawn to this conclusion by a very famous experiment of A. Michelson and E. Morley in 1887 which tried to verify that the Earth really does move through the fixed space of the Solar System at 30 kilometers (18 miles) a second. For all practical purposes "fixed space" meant the immovable, ethereal medium, *ether,* which was presumed to exist as the carrier of light waves. They split a beam of light into two parts by means of a half-silvered mirror which was part of an optical device called an *interferometer*. It was arranged that half of the light moved parallel to the Earth's orbital motion and the other half moved normally to the orbit. The light beam that travels initially *with* the Earth's motion in space and then is reflected back *against* the Earth's motion *should take longer* to traverse a given distance than the beam which moves out and back in directions normal to the orbital motion. This prediction was not verified; each light beam took the same time as the other to traverse the same distance. We cannot say, therefore, that the Earth's motion through the fixed space of the Solar System, a space in which light presumably moves at a constant speed in any direction, will influence our measurement of light's speed.

Einstein's explanation of the negative result of the Michelson–Morley experiment was that all physical phenomena, from swinging pendulums and attracting magnets to the movement of light itself, are unchanged when shifting from one constant-velocity frame to another. Starting with this postulate we are led to incredible conclusions.

For example, if you shine a searchlight from a spaceship moving at 99% of the speed of light, ob-

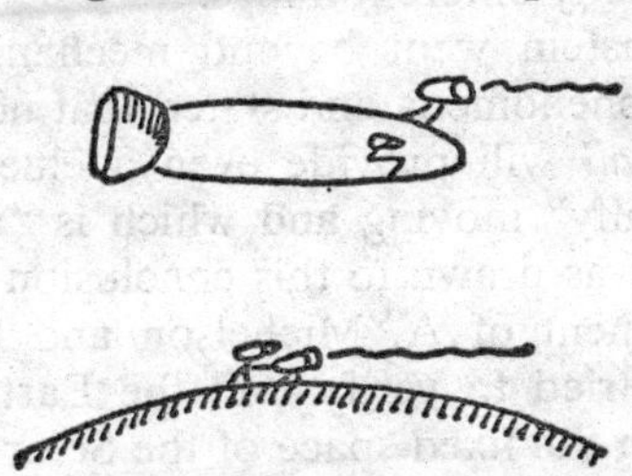

FIG. 9. The light shining from the spaceship will move through space at the same speed as the light from the Earth, as measured by both observers (atmosphere neglected).

servers *either* on the spaceship or on the Earth will measure the same value for light's speed. (Needless to say, it follows that observers on "Spaceship Earth" will measure the same speed as those on Saturn's moons.)

If the speed of a source of light were somehow to influence the speed of light, as a gun's speed influences a bullet's speed, then the light which reaches us from double stars ought to be affected by each star's velocity toward or away from us. The observed two-body motion would then depart from the laws of celestial mechanics. This, however, doesn't happen.

If, on the other hand, the light speed *in vacuo* were influenced by a real ether which acts as its carrier—just as the sound of a foghorn would reach us sooner if it were carried toward us in a strong wind—we would inevitably find ourselves in an embarrassing dilemma. Does each planet and star carry along its private ether? Where is the boundary from one to another? If the ether does have a velocity, to what absolute reference can it be referred? The Sun perhaps? Sirius? Or somewhere completely out of the Galaxy? Where will we find this fixed reference frame, which no doubt must be the real center of the Universe? We can search forever but we will not find it. Nor have careful measurements such as those of the Michelson–Morley experiment, and its many variations with light and radio waves, been able to locate it. One reference frame is indeed as good as another, and neither one is absolute. The symmetry is complete.

We must abandon that remnant from the mechanics of waves in material objects, such as sound in air, that light or any other electromagnetic radiation must be supported by some sort of medium.

Let us return to the two frames of reference described at the start of this section—the "moving" train and "stationary" station. We switch on a light in the center of a car and in the world of the traveler the wave moves *spherically out,* striking the front and rear

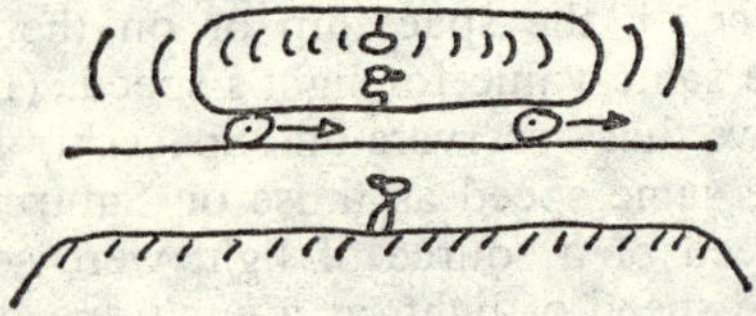

FIG. 10. The light spreads out as a spherical wave centered at the bulb, which we may assume flashes on for an instant. The speed of expansion is exactly the same for each observer. Yet, unlike the traveler, the station master sees the rear end of the car being struck by the light before the front end.

walls of the car at precisely the same instant. But what of the station master's world? *He too notes that the light moves spherically out,* uninfluenced by the motion of the light bulb which emitted it. And in his world the rear wall of the car is struck by the light first, flashing a fraction of a second before the front wall. Likewise, if the light bulb were to flash on just as the car and bulb passed the midpoint of the station, the station master would see the ends of the station lit simultaneously. This event would be hotly disputed by the traveler who "knows" that the front end of the station is struck first.

The only way we can accept this strange situation, which comes about by insisting on the constancy of the speed of light in the empty space of both reference frames, is by concluding that the light wave or quantum moves through a space and time that is *different* in one frame when observed from the other!

The equations which compare the *xyz* space variables and *t* time variable in the two frames—called the *Lorentz transformation equations*—are found very easily in the following manner: The speed of one frame relative to the other is specified, say, along the *x* axis. Then, a perfectly spherical bubble of light expanding uniformly at the *same speed* in each frame, is described in both frames by the simple equation of an expanding sphere in three dimensions and time—*four dimensions in all.* Although the two equations, one for each frame, are equivalent, the coordinates are not. They differ only in that $x' = x + vt'$, where v is the speed of the unprimed frame in the space of the primed frame. With only these facts the transformation equations relating xyz and t in terms of $x'y'z'$ and t', and vice versa, are found.

The two immediate conclusions which emerge are that all lengths in one frame parallel to the direction of motion are *foreshortened* and time moves *more slowly* as observed from the other. The station master sees the traveler light his pipe and open his newspaper, all in slow motion. The traveler notices that the station master walks in slow motion into his office. Each sees the other as unusually skinny, flattened along the x axis of train motion. The trip between cities takes less time than would be expected if one were merely to divide the known distance as stated on a map by the speed of the train, for the traveler sees the world moving at high speed past his window and so along the track all distances in the moving frame of the landscape are shorter as observed in the "stationary" frame of the train.†

Does it all happen this way? Not really. Neither the station master nor the traveler would notice anything unusual at all unless the speed of the train were not just 100 kilometers/hour but close to 300,000 kilometers/*second,* the speed of light *in vacuo.*

If the day should ever come when spaceships travel at speeds close to that of light, *distances of a few hundred light years could be spanned in only a few years,* yet *without* the vehicle ever exceeding the speed of

† At this point we can see how the negative result of the Michelson–Morley experiment was explained. The light path length in the direction of Earth's motion was slightly shortened by just the right amount to bring the reflected light beam back on schedule.

light *in vacuo,* a speed which, according to the relativity theory of Einstein, is impossible to attain.

The Einstein theory discussed here is actually called the *special* theory of relativity as it is limited to the special case of reference frames moving at *constant velocity* with respect to each other. If the frames should, on the other hand, accelerate with respect to each other—for example, a train rounding a curve, or one speeding up from rest, or a rotating space station—the *general* theory of relativity would be applicable.

There is one more important conclusion to be drawn from the special theory. Coming back to the fundamental postulate that all principles of physics, including those governing the transmission of light, are identical in both frames, we look now at that important principle of conservation of linear momentum in the absence of an applied force. The principle must retain its validity when transforming from either frame to the other. Suppose, for example, the train traveler throws a ball out of the window to the station master. In directions which are perpendicular to the train's velocity there is no contraction of length (this contraction takes place, remember, only in the forward direction). But the station master clearly sees a *time dilation* in the train, and vice versa, as we mentioned before. There is, of course, no applied force controlling the ball's motion (gravity plays no role in the argument so it is not considered here, or we must move to a planet where gravity is negligible) as it passes between train and platform. Therefore, the linear momentum transverse to the train's motion, or mass times transverse velocity, is maintained at a constant value in either frame according to Einstein's fundamental postulate. There is only one way for the station master, who has now also become a master of physics, to accept the slow motion of the mass in the direction normal to that of the train's motion and not violate the principle of linear momentum conservation in this same direction—the ball's *mass must indeed be greater* when it is virtually at rest or moving slowly in the moving reference frame and measured in the other frame. If the train were moving close to the speed of light, then all objects in it, including the thrown ball, would have enormous masses in the frame of the platform, but quite normal *rest mass* in the world of the train. Naturally, if the station master should throw a ball at the traveler, it would cross the car carrying the motion of the backward-moving station and so to the traveler what started off as a golf ball in the hand of the station master might now become a diminutive high-speed bowling ball.

What determines how massive the golf ball becomes? As with both time dilation and length contraction, the mass increase is equally sensitive to the ratio of the speed of the moving frame (e.g., the train or platform as viewed from the other) to that of light. This ratio can approach unity but could never become unity—if it did the transformation equations would require that dimensions parallel to the direction of motion would shrink to zero and all masses would become infinite! Despite the impossibility of material objects ever reaching the speed of light, it seems paradoxical that by application of the Lorentz transformation equations a vehicle could pass two points in space in less time, as measured by its own clock, than it would take light to cross these points as measured in the frame in which these points are at rest. Thus we get the illusion of traveling faster than light. It is, to be sure, nothing more than an illusion. In the spaceship's frame a light beam does cross the distance sooner than the ship, for the points are closer together in the very real world in which it is stationary and the rest of the Universe moving rapidly. In the outside world in which the points are at rest, the spaceship does take longer than the light beam for the ship, after all, can never move faster than light!

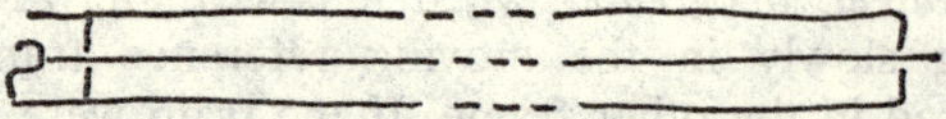

Fig. 11. The two-mile long linear accelerator of Stanford University. The electrons are under powerful forces to push apart from each other and therefore defocus the beam. However, they move so close to the speed of light that the mass of each, or resistance to force, multiplies thousands of times in agreement with Einstein's special theory of relativity. This reduces their acceleration away from the central beam axis, so that the sharpness of the beam is preserved.

In Stanford University's two-mile long linear accelerator the electron beam almost immediately reaches a speed exceedingly close to the speed of light. Yet electromagnetic energy pours into it all along its path. "Accelerator" is scarcely the proper word to describe the main effect on the electrons. They increase their mass several thousand times beyond their rest mass. Although all of the electrons are of the same electric charge and would ordinarily be expected to push apart from each other by their electrostatic repulsion and spread out, or defocus, the needle-sharp beam, this does not happen. Their mass increase curtails their tendency to acquire sufficient transverse velocity, and the sharpness of the beam is preserved throughout the two-mile length (which, to the electrons, appears no more than a few feet). This effect is an instance of Newton's law of motion in action, valid as well in relativity theory. The force of electrostatic repulsion applied to the full momentum of the electron causes its mass–transverse velocity product to change with time in responding to the force. This is the general form—the time rate of change of momentum equals the force—intended by Newton. (The familiar expression, force equals mass multiplied by its acceleration, works only for the special case when the mass is constant with time, as in most everyday dynamical problems.) The linear accelerator provides but one of numerous examples where mass is not constant with time; these examples verify, in each instance the Einstein principle of relativity.

The deduction that mass increases with speed enters into a concept of the future–nuclear-powered rockets. When the nuclei in a *critical mass*–that is, a mass large enough to sustain a *chain reaction* throughout the material–divides or fissions, and we gather together the products of the reaction, we discover an apparent violation of the classical principle of mass conservation. Almost 1% of the initial mass of uranium has vanished. This so-called *mass defect* has been converted into energy in an amount which can be found from a famous equation of Einstein drawn directly out of the above discussion: The energy released, E, is equal to the mass lost, m, times the square of the speed of light, c. That is, $E = mc^2$.

This excess mass of uranium beyond that of its *fission products* is a measure of its instability, its readiness to yield large amounts of energy in forming the stable elements comprising the final fission products of smaller total mass. But to what extent have we really lost mass in the reaction? At the instant of the fission, as the newborn reaction products as well as neutrons and other particles begin to move out at tremendous speeds, the total mass seen in the original reference frame is large, by the argument above, exactly what it was in the uranium *before* it fissioned. However, as these high-speed particles collide with the rest of the Universe and are finally brought to rest in the same original frame as the initial piece of uranium, then, of course, there is a measurable mass defect. The energy has moved on, and even some of the matter in the Universe that has been struck and swept ahead by the blinding splash of mass–energy is a bit more massive than before, until it too comes to rest. The energy released is then measured by the loss in mass that appears when all the reaction products are brought finally to rest. It arises from the rearrangement of the very strong forces that lie within the atomic nucleus. In both the nuclear explosion and the linear accelerator mass and energy are inseparable.

Angular momentum

As distinct from linear momentum (the mass–velocity product), angular momentum deals with the nature of motion about a point in space, or more correctly, *about an axis* through the point. Although it is customary to think of rotating objects such as flywheels or bicycles, when we speak of angular momentum, we may also find that the angular momentum of an object mov-

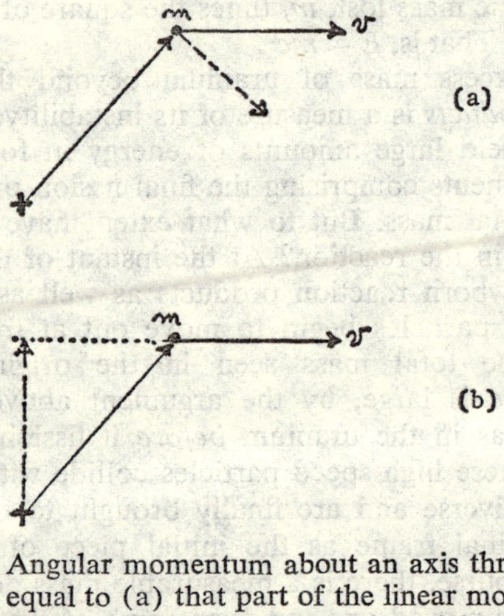

FIG. 12. Angular momentum about an axis through an origin is equal to (a) that part of the linear momentum which is normal to the direction to the origin times the distance to the origin, or (b) the component of the line from the origin that is normal to the direction of motion times the linear momentum.

ing even in a straight line has important physical significance.

At any instant the angular momentum of an object referred to some origin is *the component* of the linear momentum that is perpendicular to the direction from the origin to the object times the distance between the origin and object. It could alternatively be defined as the product of the linear momentum and that portion of the line from the origin and the object which is projected perpendicularly to the direction of the linear momentum. Either way it denotes the same tendency to turn about an axis through the origin. If we imagined a turntable at rest in space capable of rotating about that axis and suddenly held to our moving object, how fast would the turntable rotate? This would provide a direct measure of our object's angular momentum.

Suppose no forces at all were to act on our object, so that its linear momentum remained constant (i.e., straight-line motion at constant speed). Then, even though the line from the origin to the object were to grow longer and longer, the *component* of that line normal to the direction of the linear momentum always stays the same (see Figure 13), and so the angular momentum remains constant as it should.

There is another way for the angular momentum to be unchanged from some original value *even if a force were acting,* and even if its linear momentum changes to follow this force in accordance with Newton's second law of motion. That situation occurs if the force were *radial,* that is, if it acts on the object in a direction toward or away from some origin; the force is either attractive or repulsive, respectively. Angular momentum is then conserved, for there simply is no way for a radial force to pull or push the object around the origin.

Thus, every satellite moving exclusively under the influence of a radial force—a force derived either from a point mass or spherically symmetric mass—will move

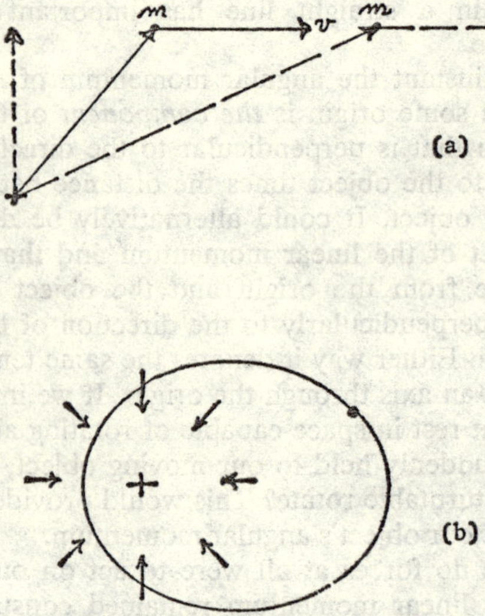

FIG. 13. (a) Angular momentum is conserved if the linear momentum is conserved, i.e., if no force is acting on the moving object. Here we see that the component of the line from the origin normal to the direction of motion never changes.

(b) The angular momentum is conserved even if a force is present, provided it is *central*—directed toward or away from some origin. In this case the angular momentum is defined with respect to this origin.

in such a way that its angular momentum about an axis through the center of symmetry *remains constant.*

Imagine an object "infinitely" far from its gravitational source and in motion by an "infinitesimal" amount. How could it possibly have a finite angular momentum? The distance approaches infinity. The velocity approaches zero. Yet the product of the two—or more to the point, the straight-line distance to the gravitational source times the component of the velocity normal to this line—is finite. Not only is it finite but it remains constant, even as the gravity field pulls it in all the way in a parabolic path. It is easy to calculate this constant at a special point—the distance of closest approach, when the straight line to the center is *exactly normal* to the direction of motion as it races around the gravity source at the highest speed it will attain. It continues on in the other arm of the parabola symmetric (with the first) about the line from the near point and the center. Then it moves on to infinity moving slower and slower with each passing second.

In the very special case when the object is exactly motionless, relative to the gravitational source and at any arbitrary distance from it, then angular momentum is not only conserved, but is conserved at the value *zero*. Its dynamics is one dimensional, radial motion all the way as it picks up speed under the steadily increasing gravitational force.

But in the general case, a very tiny nonradial motion at a very great distance to the gravity force center can account for a significant angular momentum, observed eventually as a high-speed swing about the gravitational center.

There are several relations in physics described by the finite product of two numbers, one of which is infinitely large and the other infinitesimally small. In the theory of quantum mechanics, for instance, the *Dirac delta function* plays an important role. Suppose we should require that some function of x exist precisely at x but nowhere else, and furthermore that the area under the curve of the function and the x axis be finite, say unity. This is quite difficult to perceive, for x is a point. So we move a very tiny distance out from x, and allow the function to rise to only so high a value as to make the area of the narrow rectangle the required finite amount (see Figure 14). If the displacement from x were made half as great, then the function would be twice as high to keep the area at its former

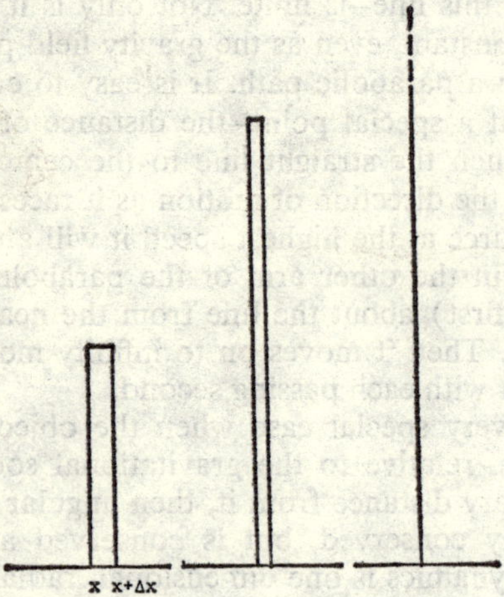

FIG. 14. The Dirac Delta Function. The *integral* of the function through the entire range of x is unity, i.e., the area under the curve is unity. Yet the function is zero everywhere except at x. So it must shoot up to infinity as the displacement from x becomes infinitesimal. The product of the base and height of the "rectangle" still is unity.

value. Finally, if the displacement from x were infinitesimally small, the function would now have to soar to an infinitely high value, the very same kind of thing we encounter when we speak of angular momentum at an infinitely great distance.‡

Earth satellites will almost, but not quite, follow this rule of angular momentum conservation, for the

‡ The temptation is great for a science-minded songwriter to come up with "Our Love is Like a Dirac Delta Function" (". . . The intensity to infinity so it ran/But why, tell me why, was it over just when it began?"). The honors for introducing it ought then to go to the granddaughter of Nobel laureate Max Born, one of the early pioneers of quantum theory—Olivia Newton-John.

Earth's gravity is almost, but not quite, radially directed. In any event there are always disturbing gravity fields at the satellite which originate from the Moon, Sun, and nearby planets. These *perturb* the satellite's motion and distort its ellipse.

A rocket under a small thrust but in orbit about a pure radial force field, which we call a *central field,* will deviate very slightly from constant angular momentum motion by an amount depending upon the magnitude of the thrust compared with the central force (e.g., it might be 1% of gravity, or 0.1%, or less).

If a satellite were to encounter air resistance or meteoroidal impact, this too would prevent angular momentum from being conserved, for now an additional force beyond the central gravity force controls the satellite's motion. Also, a satellite with iron parts or with electric circuits which produce magnetic fields would be disturbed by the Earth's magnetic field, as well as the magnetic field of the Sun which reaches out and is enhanced by streams of charged particles, called the *solar wind.* The net result of any or all of these *perturbing forces* would be a distortion of its elliptic orbit.

Orbital energy

The principle of angular momentum conservation of an object moving exclusively under a central force field, such as the radial gravity field of an isolated spherical planet, is one of the two basic laws of *celestial mechanics*. The other is conservation of mechanical energy. Together they describe the past and future history—location and velocity—of objects subject only to gravity fields. When we study the motion of a rocket *while* its engine is firing or *while* atmospheric drag is trying to take over, we are in the realm of general *space flight mechanics* and the analysis of the motion becomes more complex. Celestial mechanics becomes complex enough when you have to account for multiple moving sources of gravity, or when we are faced with distortions caused by a gravity source (like the Earth) resembling a pumpkin more than it does a basketball, or when a satellite we are studying is massive enough (such as the Moon) to grab hold of the gravity source (in this case, the Earth), which controls its own motion and move it about.

But we must start from basics—an artificial satellite, perhaps, moving about the Earth (which is assumed to be spherical), or sent on its way effectively out of the Earth's field to the Moon or Mars. In Chapter Two it was easy to find the escape speed needed to launch an object from the surface of a planet (at a distance r_0 from its center) out to a region of space where the gravitational force field is negligibly small. It was

$$\sqrt{2g_0r_0},$$

where g_0 is the gravity force field at the surface. Air resistance is neglected. This speed is independent of the direction of launch. If the planet were rotating this same launch speed would be needed, but a portion of it would be contributed or withdrawn by the motion of the planet's surface in space caused by its axial rotation. This is one reason why so many Earth satellites are launched from Cape Canaveral. They rise up vertically and then bend over horizontally toward the east. If they were launched at the equator the Earth would contribute 460 meters/second to the launch velocity, that is, just over 4% of the required escape velocity or just under 6% of a low orbital speed. The effect is somewhat less at the latitude of Florida.

In a horizontal launch at the escape speed, the spacecraft is given the maximum angular momentum possible for a given launch speed as its launch direction is exactly normal to the line to the Earth's center. It rises higher and higher as it swings in a parabolic curve that moves off to infinity. The angular momentum which it had at the start,

$$r_0 m v_e ,$$

where m is the spacecraft's mass, is retained throughout its motion. The speed gradually approaches zero, yet the perpendicular component from the planet's center to the extension line of the direction of motion gradually rises without limit. Nevertheless, as we noted earlier, the product of this very large and very small number is the same as it was at the start.

What about the energy of the spacecraft? At launch it had its greatest speed and therefore greatest energy of motion, called *kinetic energy*. As it rose to greater and greater heights, it gained energy of position, called *potential energy,* but in doing so it moved more and more slowly, continuously losing its kinetic energy. Potential energy is the required amount of work needed to bring an object from one point in space to another in the presence of a force field such as gravity. For example, if we raise a sack of potatoes from the floor to a shelf a certain amount of work must be done, but if the shelf were twice as high then twice the work is needed to get it there. If it fell from the higher shelf it would strike the floor with twice as much kinetic energy (though not twice the speed) as from the lower shelf.

Because a launched object's kinetic energy loss is made up by an equal amount of potential energy gained in rising, the *total* energy–potential plus kinetic –is constant at every point in the projectile's path. Conversely, in falling the kinetic energy gain is made up by an equal amount of potential energy loss. If we set the zero reference level of potential energy to be at the Earth's surface so that at an "infinite" distance from the Earth the potential energy level is an amount $g_0 r_0$ above the Earth's surface (see p. 18), we can write the total energy of the launched one-kilogram object as:

kinetic plus potential energy at launch
= kinetic plus potential energy at infinity,

$$\tfrac{1}{2}v_e^2 + 0 = 0 + g_0 r_0 .$$

Here we expressed the total energy at the two extreme points in the object's path to obtain the familiar expression for v_e in terms of g_0 and r_0, repeating the result which we found in Chapter Two.

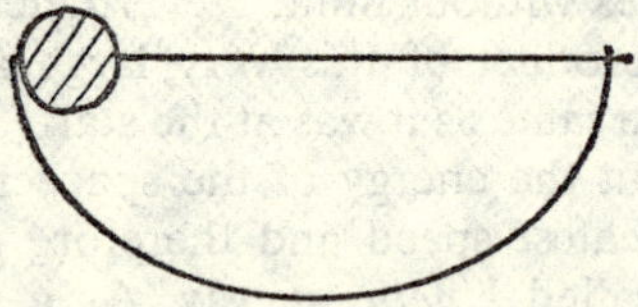

FIG. 15(a). For the same launch speed the apogee of the vertical launch is higher, as all of the kinetic energy of motion is converted into potential energy of location. In the horizontal launch, however, part of the original kinetic energy remains at apogee.

Now suppose we launched our kilogram vertically and at a speed *less* than v_e, so that it reached a high point, called *apogee,* where it lost all its kinetic energy, and then fell straight down to Earth again. This corresponds to Figure 7 and the discussion in Chapter Two.* The maximum height can be calculated in terms of the launch speed, and *vice versa.*

The problem of finding the maximum height attained by the *horizontally* launched projectile, now moving in an ellipse, is slightly more complicated. This is because the projectile here has angular momentum. At apogee

* We use r_a for the apogee distance from the center, so that the maximum height is $r_a - r_0$ from the Earth's surface. The launch speed is v_0. The energy equation corresponding to the one above, but where the right-hand side expresses the kinetic and potential energy at apogee is,

$$\tfrac{1}{2}v_0^2 + 0 = 0 + g_0 r_0 (1 - \frac{r_0}{r_a}).$$

in the ellipse the projectile had finite speed and kinetic energy while in the straight line it had none. Therefore, for the same launch speed, v_0, its residual speed at apogee prevents it from rising to quite as high an apogee point as in the vertical case.†

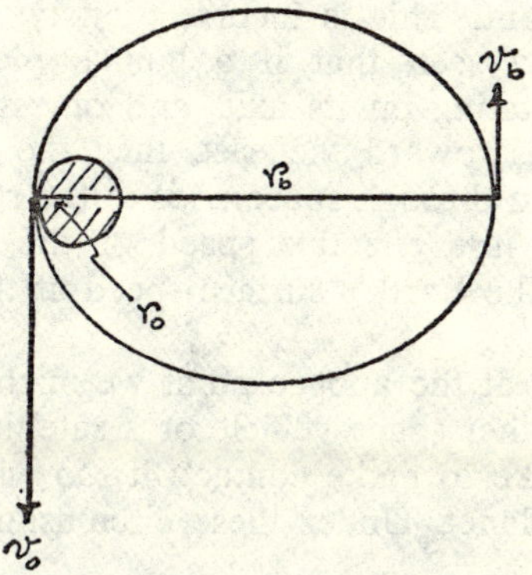

FIG. 15(b). The constant angular momentum is most easily found at the two extreme ends of the elliptical orbit, for here the direction of motion is exactly perpendicular to the line from the origin of the gravitational force field. Thus we have the simple relation: $v_0 r_0 = v_b r_b$.

† We call the apogee distance in the ellipse r_b. Both r_b and the speed at apogee, v_b, can be found from the energy conservation equation *together with* the angular momentum conservation equation. The energy conservation equation is

$$\tfrac{1}{2}v_0^2 + 0 = \tfrac{1}{2}v_b^2 + g_0 r_0 \left(1 - \frac{r_0}{r_b}\right).$$

The angular momentum conservation equation is simply

$$v_0 r_0 = v_b r_b\,.$$

As in the energy conservation equation, values are taken at the two extreme points in the orbit–the launch, or *perigee* point, which is the nearest point of the ellipse to the center of the Earth, and the apogee point.

Rendezvous

One might get the mistaken impression—because the height attained in the vertical launch is greater than in the horizontal one, even though the launch speeds are the same—that it is more economical to carry out a space mission with a vertical launch, and thereby use less propellant. This is incorrect on two counts: We have already seen that if a launch planet, say the Earth, is rotating on its axis, and our space vehicle's flight path is toward the east, then the Earth's rotational motion through space adds to the rocket's speed. In this way the effective speed of the rocket is increased by the Earth's surface speed at the latitude of launch.

Secondly, at the apogee point we might wish to encounter another space vehicle or a satellite with which we would like to make contact; it too will be orbiting about the planet. Under these circumstances it would

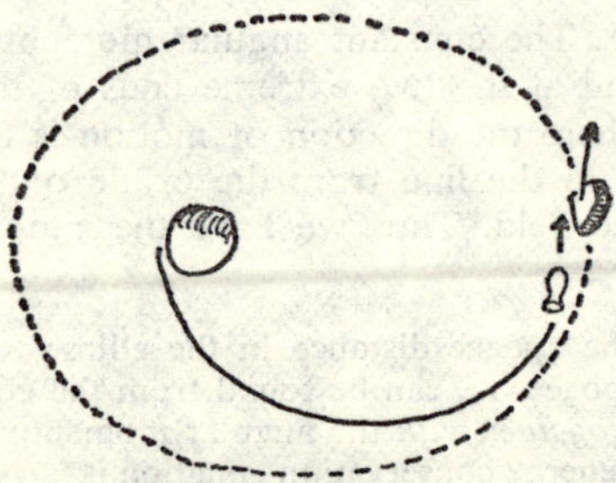

FIG. 16. Closely matched speeds between a space platform in orbit and the spacecraft at apogee of its ellipse.

be far easier to link up with it at apogee if both our own vehicle and the target are moving through space *in the same direction* at the apogee point of our flight path. Therefore we would wish to launch horizontally in such a direction that our velocity at apogee will be parallel to that of the satellite with which we wish to make contact. We will discover in Chapter Seven that our spacecraft will never reach apogee with a speed greater than the satellite's; our speed will always be less (unless the satellite is in the identical orbit) so that at apogee we must activate our rocket engine briefly in order to match speeds with the satellite. The vertically launched rocket, on the other hand, would arrive at its apogee with zero speed; in order that a linkup take place, it would be necessary to activate its rocket engine with sufficient thrust to match the full speed of the satellite. Although at such an altitude this would be much less than an orbital speed close to the Earth, the large inventory of propellant needed to perform this mission still must be carried aloft. In summary, for a given launch speed, and energy, the horizontal flight path will give us the greatest possible angular momentum which will simplify the problem of matching speeds with another orbiting vehicle moving in the same direction, or, for that matter, merely changing to a larger orbit.

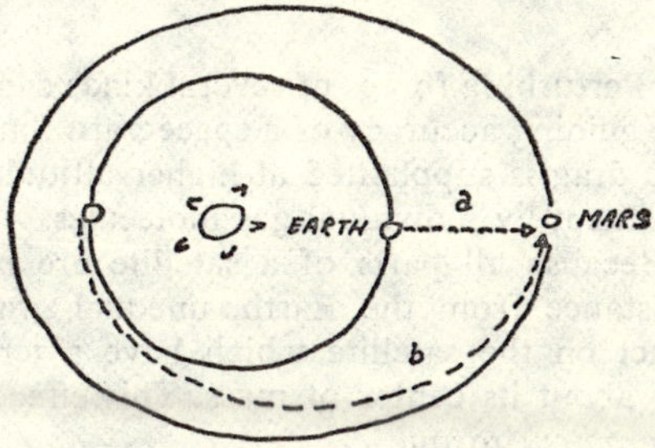

FIG. 17. Alternative flight paths from the Earth to Mars. All planets move about the Sun in the same direction. Far more propellant is conserved by making full use of the Earth's orbital speed and by reaching Mars with closely matched speeds for either a soft landing or entering a Mars satellite orbit. Therefore flight path *b*, that of greatest angular momentum about the Sun, is chosen.

While we are on the subject, let us look at the problem of sending a planetary probe from the Earth to Mars. The two planets are in orbit about the Sun, in a fairly circular ellipse in the case of Earth and a slightly

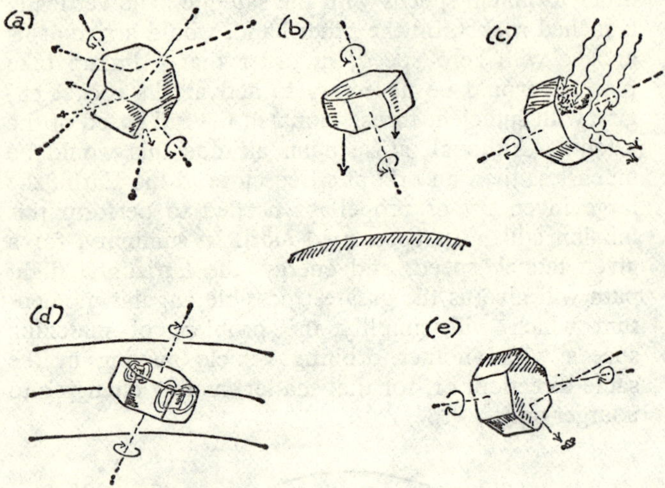

FIG. 18. Perturbing forces of several kinds can interfere with the guiding accuracy of a spacecraft. Simple aerodynamic drag is supplanted at higher altitudes (a) by bombardment by individual gas molecules.

(b) Because all parts of a satellite are not at the same distance from the Earth, unequal gravitational forces act on the satellite which have a tendency to rotate it about its center of mass. This effect is called *gravity-gradient torque*.

(c) A radiation torque can occur if different parts of the spacecraft reflect or absorb sunlight unequally.

(d) A magnetic torque is produced by the interaction of current loops inside the satellite with the Earth's magnetic field.

(e) Meteoroid bombardment can also upset the guidance.

eccentric ellipse in the case of Mars. It would be madness to choose a "radial" flight path, the shortest line, from Earth to Mars, for by launching instead in the direction of the Earth's 30 kilometers/second (18 miles/second) orbit about the Sun the rocket moves off toward the orbit of Mars with minimal propellant expenditure having taken full advantage of the kinetic energy on the launching pad in the solar-based reference frame. Furthermore, it reaches Mars with sufficient speed to almost match that of the planet.

CHAPTER FOUR

The Moon as Ion Rocket . . . and Other Stories

Space cannon

"Beware of the concussion!" The shout from the giant loudspeakers rang across the valley back and forth over the heads of the running mob. They poured into the cliff-rimmed field like an army of giant ants—armed with clubs, sticks, rocks, anything they could lay their hands on. "Beware of the concussion," it echoed, ". . . the concussion, the concussion, the concussion." The mob converged on the five lanes which radiated from the valley center like the spokes of a colossal wheel. The towering space cannon rose from the hub in the red dawning sun to cast a shadow that had no end. The space capsule was lowered into the gun barrel. A dream of ten thousand years was about to become reality as humankind prepared to step off his home planet for the first time, toward the Moon.

"We'll hate you more if you succeed than if you fail," shouted the leader of the mob. "Stop this insane progress," his own loudspeaker focused toward the high cliffs where the masters of the Technetronic Age were programing the coming event. "Where does it bring us? What measure of happiness?" But it was too late. It was all over. The button had been pressed. The capsule turned into a silver ray of light among the morning stars, and at last it faded from view.

Thus, H. G. Wells's prophetic film odyssey through

a hundred years of history beginning with the 1930's, *Things to Come,* drew to its dramatic ending. The first prediction, the start of World War II on Christmas Eve 1939, was accurate to within four months; it was the first of many wars, followed by pestilence and return to cave dwelling. Society was completely disorganized and dominated by scores of petty dictators such as the one portrayed by a youthful Ralph Richardson. The technological renaissance and emerging Age of Reason was personified by a Lincolnesque Raymond Massey. Then at last, by the mid-thirties of the 21st century came the inevitable revolt of the artisans. As in Francis Bacon's *New Atlantis,* the building of Utopia had been entrusted to the technologists, but the disturbing effects of their handiwork on the quality of life was not to be easily accepted.

The British Interplanetary Society, while vigorously applauding the film, was puzzled by at least one technical question: How were the two astronauts in the capsule (a man and a woman–lovers of course) able to withstand the terrible acceleration stresses that must have taken place within the gun barrel to enable them to reach close to the escape speed at the muzzle? The more curious members of the Society also wondered how the astronaut pair could have landed on the Moon, or how they could have gotten back, or even how the capsule could ever have held together in the Earth's lower atmosphere at the fantastic speed which was needed to get it to the Moon. The film ended just in time to avoid these questions.

Even the prophetic Wells dared not predict that space flight fiction would become fact only three decades after the film was made, with rockets exchanged for the space cannon and its capsule. Was Wells really guilty of a technical *faux pas* in choosing a space cannon for the space voyage?

The human body and delicate instruments are both limited in the amount of acceleration stress which they can tolerate. For a few seconds, or even minutes, in vertical flight accelerations of nine times gravity may be withstood, but this is just about the upper limit. As we will see in Chapter Six, under these conditions the stress on an astronaut will then be equivalent to ten times his or her normal weight. What length space cannon should then be designed in order to curb the buildup of velocity with time so that the stress lies below this tolerable limit? Equal to the height of a 20-story building? A 100-story building? Here is one place where it pays—borrowing the phraseology of the engineer—to put numbers into the equation.

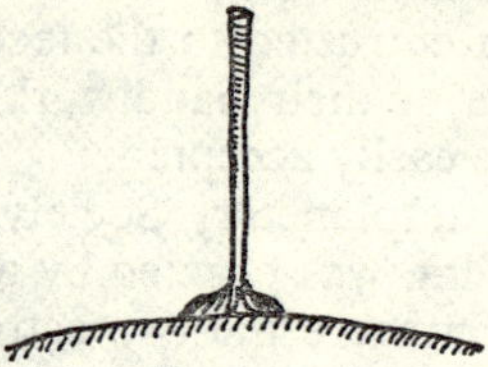

FIG. 19. The space cannon of H. G. Wells or Jules Verne would have to be one-ninth of the Earth's radius —or *700 kilometers*—in length in order to insure that the astronauts in the capsule inside will reach the speed needed to escape Earth's gravity at the top and still not encounter stresses greater than ten times normal weight throughout the time interval of acceleration within the gun barrel.

We know that if a projectile starts from rest at one end of a gun barrel of length l and is constantly accelerated by an amount a to a muzzle velocity v_e, then*

$$l = \frac{v_e^2}{2a}.$$

If the muzzle velocity were the escape velocity, then we can express v_e^2 as $2g_0r_0$, where g_0 is the Earth's

* This equation is found by combining the well-known relation, $l = \frac{1}{2}at^2$ with $v_e = at$ and eliminating t, the time.

gravity force field strength at the gun, or the Earth's surface, and r_0 is the Earth's radius. Furthermore, we agree to limit a to a maximum value of nine times g_0 (We note that the force of gravity on a unit mass of one kilogram has the same units as the acceleration of a body due to that force). We conclude, therefore, that

$$l = \tfrac{1}{9} r_0.$$

The cannon would have to be 700 kilometers (435 miles) long!

How did Wells avoid getting himself involved in constructing a gun of this magnitude for his space venture? Perhaps his own special scheme for launching space vehicles had not reached the ears of the British Interplanetary Society. Was the space capsule made of "cavorite," that mysterious antimatter in the Wellsian fantasy *The First Men in the Moon* that falls *up* instead of down? Hardly likely, for then there would have been no concussion to beware of. Perhaps Wells found a way to accelerate both the capsule and its contents by a new kind of *body force,* one that acts directly upon every molecule and every subatomic particle and precludes the squashing down of one molecule upon another to produce the often uncomfortable feeling which we call weight. Were such a force present, the occupants would no more feel their high acceleration than if they were falling through the sky of Jupiter or some giant dead sun itself. But if Wells's fervent imagination held any of these ideas, he kept them to himself as did his predecessor Jules Verne, who, like Wells, also launched a capsule to the Moon by means of a space cannon.

Space rocket

From the discussion above it is clear why we send up astronauts by rocket instead of space cannon. A rocket is, in all respects, a recoiling gun firing back a high-speed stream of *gas* instead of individual bullets. If we did not acknowledge the principle of recoil—Newton's third law of motion—we would be violating Newton's second law as extended to a system of masses —gun plus bullets or rocket plus exhaust. The argument is as follows: If the gun does not recoil then the center of mass of the system, on firing the first bullet, would begin moving, and therefore accelerate in the direction of the bullets even though no applied external force acts on the system. The center of mass can either move at a constant speed or be at rest with respect to the Earth, before *and after* the act of firing. So the gun must recoil in a direction opposite to that of the bullets by just the right amount in order that the center of mass meets this requirement.

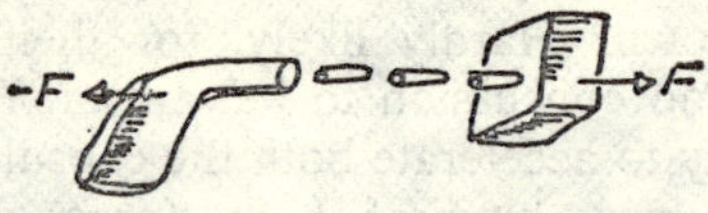

FIG. 20. Bullets imparting a force to a block of wood which absorbs them. The force is equal to the number of bullets striking per second times the mass per bullet times their speed of impact. This is the linear momentum transferred per second to the block. The gun recoils by the same force, but one which is directed opposite to that of the bullets. If the gun is not held it will accelerate in the direction of this recoil force, limited in magnitude by its mass. Apart from the downward pull of gravity and the friction of air, the gun-plus-bullets seen as a many-particle system will have *zero* net acceleration.

What determines the amount of recoil force? Like any force it is determined by the *change of momentum per unit time* which occurs, in this case, as the momentum expelled through the gun's muzzle by the bullets. The more bullets fired each second the greater the recoil force, the greater the mass of each bullet the greater the force, and the higher the bullet's velocity the greater the force. If all the bullets were stopped in a block of wood, the product, *total bullet mass striking per second* × *the bullet speed,* would be the force transmitted to the wood, and *this same force but in the opposite direction is transmitted to the gun.*

If the gun were called a rocket, this recoil force is called *thrust.* The velocity attained by a rocket caused by the action of a sustained thrust force depends, of course, on its mass, for the acceleration equals the force divided by the mass. If the mass of the rocket were assumed to remain constant during the thrusting period the calculation of velocity would be especially simple: We find the thrust as propellant mass leaving the rocket per second × its speed relative to the rocket, and divide this by the rocket's mass to get the rocket's acceleration. Then merely by multiplying this acceleration by the duration of the thrust we get the speed accountable to this sustained thrust.

But have we the right to assume that the mass of the rocket really does remain constant during the thrusting period? We know that it gets smaller and smaller every second as propellant leaves. It is, in practice, difficult to imagine a situation where we can neglect the propellant mass in relation to the rocket in solving problems. This is almost the case if the propellant were a stream of charged particles–ions–provided the thrust period were only a few hours or at most days. In an ion rocket engine the speeds of the charged particles are within several percent of that of light itself so the main factor influencing the thrust is this speed, for the mass of ionic matter leaving the rocket per second is minuscule.

Over an appreciable thrust time–months in the case of an ion rocket, a few seconds in the case of a chemical rocket–when a significant portion of total rocket mass has entered the exhaust stream, the calculation of rocket velocity must take this diminishing rocket mass into account. The result can be written in a very simple form: The rocket velocity attained at the end of a given thrust period is *directly proportional to the speed of the rocket exhaust* (as measured from the rocket) *and also proportional to the "logarithm"*† *of the ratio of the mass of the rocket before thrusting to that after thrusting.*‡

Let us look at the significance of this expression. If the mass of the fully-loaded rocket were a hundred times greater than the depleted rocket, this effect on the rocket velocity would be proportional to the log-

† The logarithm of a number, to the base 10, is the number of times ten must be multiplied by itself to produce the number. It is easy to see, therefore, that the logarithm of 10 is 1, and that of 100 is 2, but not so easy to see that the logarithm of 50 is 1.7. When numbers multiply, their logarithms add, so we find that the logarithms $2 + 1.7 = 3.7$ ought to be the logarithm of the number 5000, which it is. We need not be troubled by the meaning behind multiplying ten by itself 1.7 times to produce 50. What really matters is the addition rule, e.g., we can find the logarithm of 2 just by remembering that 50 times 2 is 100.

‡ This statement infers that the final velocity attained is independent of the thrust program—e.g., whether the propellant is shot out all at once or a little at a time. This is true so long as the basic equation of the rocket—which determines the final velocity—is valid for all cases. The basic equation considers that one molecule of propellant shooting out at some velocity relative to the rocket will cause its large mass to recoil by a small amount. The distinction between the molecule's velocity relative to the rocket alone and the rocket-plus-molecule's center of mass is ignored (it is considered a "second-order effect" in describing the dynamics). The dynamics are quite different for bullets shot out of a gun, however; these are rigid bodies, billions of molecules tied together. The gun's recoil velocity with each firing is determined by the mass of each bullet times its velocity *relative to the center of mass* of the bullet-plus-gun (containing the remaining bullets). The velocity of a bullet calculated this way is different for each firing; this was not the case for the gas molecules, whether the first or last to be expelled. As a result, when the bullets are fired one at a time the final velocity attained by the gun in free space is actually significantly greater than it would be if all the bullets could be fired at once!

arithm (to the base 10) of 100. Now 100 can be written as 10^2 which makes it easy to see that its logarithm is 2. Yet if the "mass ratio" were instead ten times greater than this, or 1000, or 10^3, its logarithm is 3. Thus, even though the mass ratio was increased ten times, its effect on the final rocket velocity is an increase of but *50%*. On the other hand, a tenfold increase of exhaust velocity results in an increase of *ten times* in rocket velocity attained.

And so, although it is true that mass-ejection rate and exhaust velocity play equal roles in affecting the thrust, the far more important factor is exhaust velocity when determining the practical effect of the thrust in the attainment of high rocket speeds after a given "burning time." By keeping exhaust velocity as high as possible, maximum use is made of every gram of propellant until all the fuel is used up.

The rocket problem has not included external forces such as aerodynamic drag and gravity, both of which could diminish the final rocket velocity for a given mass ratio and propellant speed. Gravity would not, however, reduce the rocket speed accountable to thrust during the burning period, provided the thrust force were set horizontally.

Ion and chemical rockets

The need for high exhaust speeds in the chemical rocket is so great that the combustion chamber must be kept at the highest temperature feasible. Also the lightest element of all, hydrogen, is the prime constituent in the exhaust, for the lighter the particle the faster is its own speed in a high-temperature environment.

In an ion rocket the problem is quite different. Despite the high beam speeds—almost ten thousand times greater than in the exhaust of a chemical rocket—the thrust is a tiny fraction of the rocket's weight for so lit-

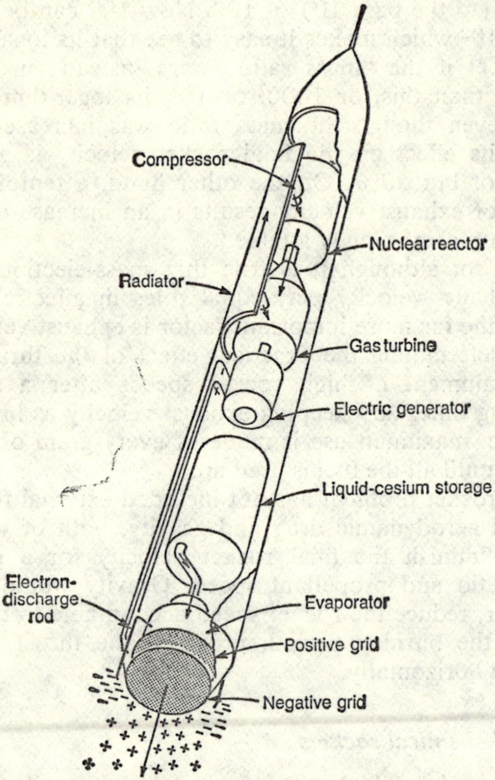

FIG. 21. A rocket propelled by accelerated cesium positive ions. Electrons also are discharged to neutralize the electric charge of the beam and thus to prevent a cloud of positive ions from forming, called *space charge,* which would reduce the propellant velocity. (From "Nuclear Rockets," J. Newgard and M. Levoy, © 1959, *Scientific American.*)

tle mass is expelled per second. It couldn't possibly lift a rocket off its launching pad. However, when applied

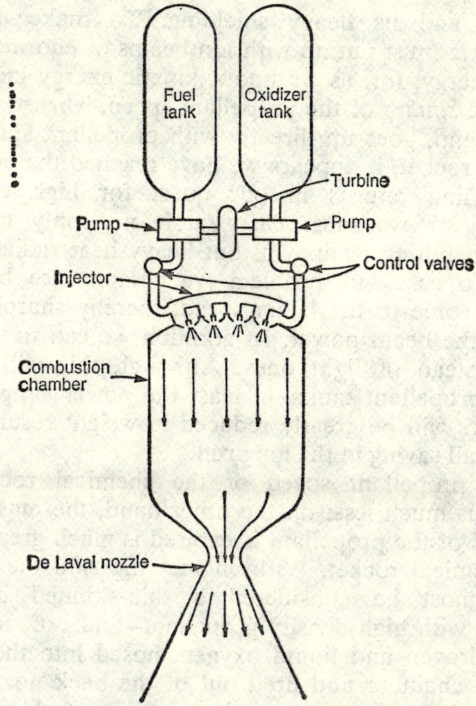

FIG. 22. A liquid-propellant rocket. The fuel could be liquid hydrogen, which is pumped into the combustion chamber along with liquid oxygen.

to a spacecraft in orbit—that is, in a weightless condition—or otherwise freely falling, a small thrust applied for a very *long time* can bring a spacecraft eventually to a high velocity and out to a great distance against an opposing gravitational force field.

One factor which limits the thrust, as well as the mass ratio which is attainable even after all the propellant is depleted, is the massive power-producing equipment which must be carried along. These are the magnets, electric generator, and possibly even a nuclear reactor and its heavy shielding. To make matters worse, we must put up with ion beams of enormous kinetic energy, for, as we know, kinetic energy increases with the *square* of the propellant speed. Thrust, on the other hand, goes up directly with propellant speed. In the ion rocket, it appears we have reached the point of diminishing returns in our quest for high exhaust speeds, for we must carry aloft not only massive power-handling equipment but heavy heat radiators as well. To ease our problem we compromise by sacrificing some of the beam speed, thereby sharply curtailing the beam power. In addition we can use heavy ions instead of light ones. Although this will mean larger propellant tanks, at least the power supply and radiators will be greatly reduced in weight resulting in an overall saving in the long run.

The propellant speed of the chemical rocket of course is much less; on the other hand, the mass ratio after all of the propellant is expired is much greater for the chemical rocket. With engineering skill the rocket can almost be considered a thin-skinned balloon bloated with high-density propellant—tanks of, say, liquid hydrogen and liquid oxygen, hosed into the combustion chamber and fired out of the back nozzle. Of course, the many pumps, pipes, and controls take up space and weight; nevertheless, mass ratios of 10 to 15 are quite usual.

As a practical example of the possibilities with a chemical rocket, let us assume that we have one such rocket on the launching pad, weighing one metric ton, or 1000 kilograms (2200 pounds). If we limit its acceleration under thrust to $9g_0$ (and thrust the rocket horizontally to avoid the so-called gravity losses which were noted at the close of the last subsection), we calculate the flight time under thrust by dividing the required speed by the acceleration. If the escape speed is desired, 11.2 kilometer/second (7.0 miles/second), dividing by $9g_0$, or 0.088 kilometers/second^2, gives a thrust period of 127 seconds. The length of the trajectory during propellant burnout is exactly the same as for the space cannon, 700 kilometers (435 miles).

If propellant were discharged at the rate of 8 kilograms/second, on the average, then 912 kilograms would be expelled leaving the remaining 88 kilograms to comprise the rocket shell, propellant tanks, engine, and payload. The mass ratio is thus 1000 divided by 88, or 11.36. We can solve for the propellant speed necessary to bring the rocket to the escape speed by dividing the escape speed by the logarithm of the mass ratio to get a propellant speed of 4.6 kilometers/second (2.9 miles/second). (A factor of 2.3 is entered into the calculation in the denominator to permit us to work with the logarithm to the base 10.) It turns out that these figures are not unrealistic for a highly efficient rocket. The likelihood of producing a sustained acceleration of this magnitude would increase, however, if the size, weight, and propellant expulsion rate were scaled up.

But what is clearly evident is that the reaction thrust of a rocket is capable of accelerating a space capsule to the escape speed *without* the need for building a space cannon 700 kilometers long.

To summarize these comparisons between the chemical and ion rockets: The chemical rocket expels nearly all of its mass at high speed in a few seconds in order to produce sufficient thrust to lift the remaining mass against gravity and into orbit. The ion rocket expels only an exceedingly small fraction of its mass per second at an extremely high speed to produce a very small thrust. It continues this expulsion over many months; in this way the rocket will have gained a very high speed, or potential energy in a gravity field, or both, despite the small thrust, yet with only a modest loss in total rocket mass. If the ion rocket were held in a gravity field the ion thrust could not overcome the rocket's support; it could at best succeed in merely lessening the support by a barely perceptible amount. Therefore, the ion beam can contribute an acceleration component from its reaction thrust only if the rocket is not held in place as when it is either in free fall (i.e., in orbit) or moving under the influence of other thrust forces or drag.

Rocket staging

We have to temper the theoretical analysis of the chemical rocket above with a few practical considerations. In the first place *gravity losses* will inevitably prevent the rocket from attaining its full 11.2 kilometer/second escape speed at the end of the 2-minute burning time. It is clear that merely to stand still above the launching pad requires a thrust equal and opposite to the vehicle's weight. Any additional thrust beyond this will send the rocket speeding on its way. The discussion, as noted above, assumes the idealization of a rocket thrusting horizontally, or in gravity-free space.

In vertical flight against gravity the mass of unburned propellant must be accelerated to higher and higher altitudes by each kilogram of propellant fired earlier through the nozzle. If the thrust force is in the horizontal direction, however, this problem does not arise, as mentioned earlier. Eliminating the gravity loss, therefore, provides an additional stimulus–beyond the reasons given at the close of Chapter Three–for establishing a horizontal flight program rather than a vertical one and forming an arc about the Earth in rising to higher and higher altitudes.

A horizontal flight program, on the other hand, does tend to aggravate an even more serious problem. Friction with the atmosphere, called the *drag force,* is proportional roughly to both air density and the square of the rocket's speed. From this viewpoint we have little choice but to rise by the shortest path possible through the dense lower atmosphere and then arc over only after the vehicle enters the thin upper atmosphere during the remaining burning time of the thrust program. This procedure will keep unburned propellant at as low an altitude as possible.

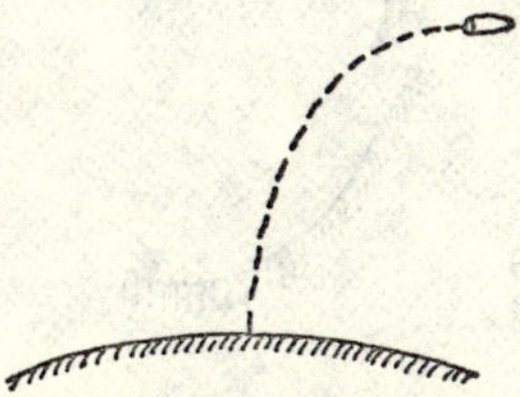

FIG. 23. A *synergic* or combined-action rocket flight path. It attempts to minimize losses caused by (1) the pull of gravity during the thrust period (i.e., losses caused by pushing not-yet-burned propellant to high altitudes), and (2) atmospheric drag. The trajectory is chosen so that kinetic energy losses which are caused by gravity equal that caused by air resistance along each increment of path. The rocket cuts almost vertically through the dense atmosphere and gradually bends over into horizontal flight as the atmosphere thins out. By moving toward the east, full advantage can be taken of the Earth's rotational speed in adding to the burnout speed of the rocket and giving it an added boost into space.

When the totality of gravity and drag losses are considered, we conclude that our actual rocket offers little practical hope of reaching the escape speed with the chemical propellants which are available. There is, however, one way out: Do not begin firing a rocket until after it has already been boosted to a high speed. Propel it down a mountainside into a valley, goes one fanciful suggestion, and fire the rocket near the bottom on the way up on the other side. A more practicable alternative has been chosen as one way to solve the problem–let the entire rocket be the *payload* of a much larger rocket, which boosts it into the thin upper atmosphere at high speed ready for a fresh launch. We can go a step further and have as payload of this newly launched rocket a small third rocket continuing on in space. In principle, we can do this

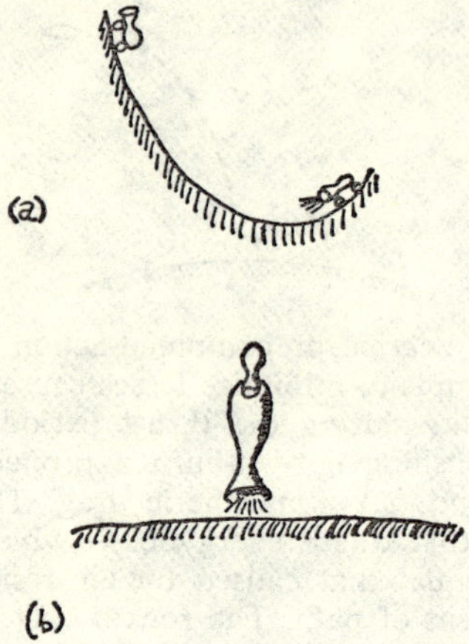

FIG. 24. (a) One way to launch a rocket, in principle, is to roll it down a mountainside on tracks and launch it at its lowest point just when it is moving at its highest speed.

(b) Another way is to make it the payload of a much larger rocket, called a booster, and again fire it when it is moving at its highest speed.

indefinitely, each new rocket "mounted upon the shoulders of its booster," as the expired and unwanted hulls, tanks, pipes, and pumps are all jettisoned before each new stage is launched. The first successful demonstration of the staging principle was made in 1949 when a "WAC Corporal" rocket, boosted by a V-2, reached the record height of 242 miles. Since then, we have come a long way.

Although it solves the problem of the attainment of high speed and therefore great distance, the staging principle has two inherent drawbacks which we ought

The earth as seen from Apollo 17.

This view of Venus from 420,000 kilometers was made by Mariner 10's television cameras one day after the spacecraft flew past the planet en route to Mercury.

Mercury from Mariner 10. (*IPS photo*)

Mars constructed from 1,500 pictures taken by Mariner 9. (*Photo from IPS*)

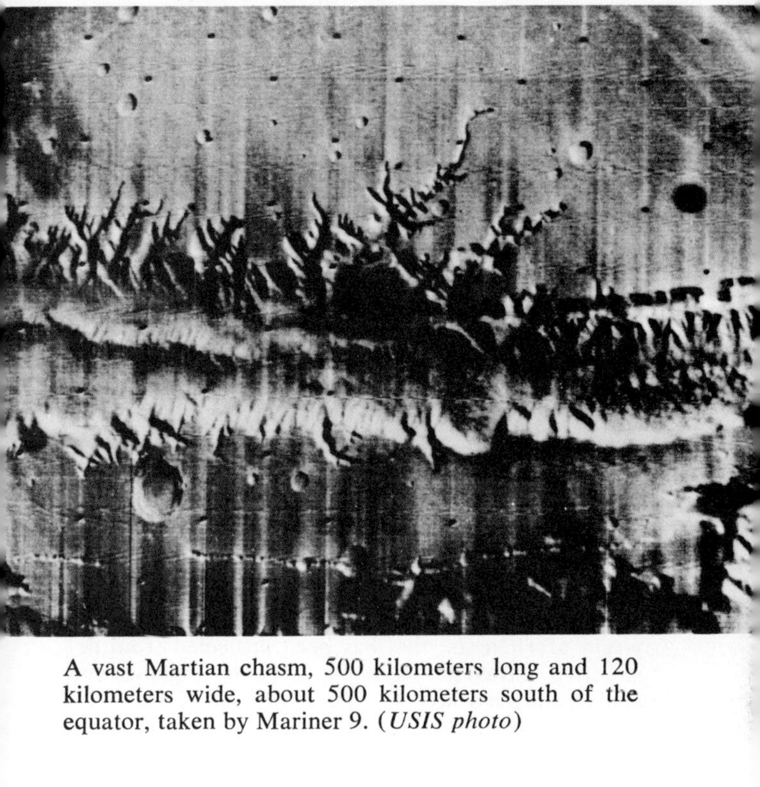

A vast Martian chasm, 500 kilometers long and 120 kilometers wide, about 500 kilometers south of the equator, taken by Mariner 9. (*USIS photo*)

This picture of the Martian landscape by the Viking 1 Lander shows a dune field with features remarkably similar to many seen in the deserts of Earth. (*National Aeronautics and Space Administration photo*)

The long arm of Viking 2 moves a rock and exposes virgin Martian soil that has been protected from destructive ultraviolet sunlight prior to digging and analysis in the search for life. (*National Aeronautics and Space Administration photo*)

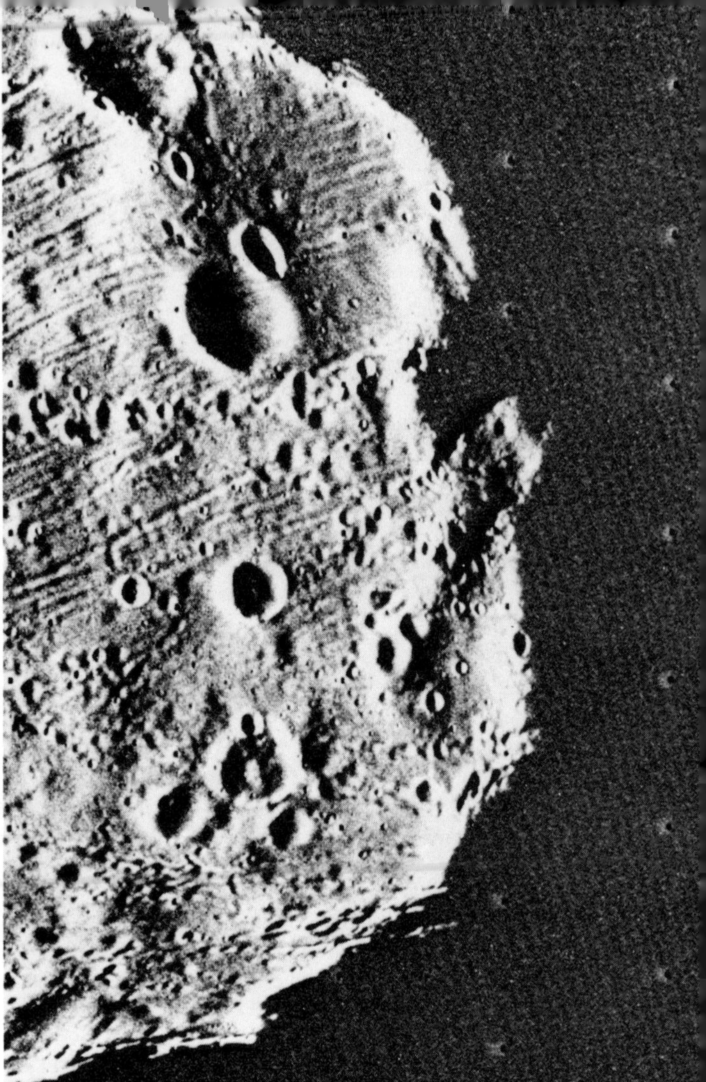

A close-up view of Phobos taken by the Viking 2 orbiter from a distance of 900 kilometers. Objects 40 meters long are resolved. The area visible here is 9 by 18 kilometers. The cause of the regular markings is uncertain. (*National Aeronautics and Space Administration photo*)

not fail to keep in mind. One is the magnified factor of unreliability. For example, if each rocket stage in a five-stage rocket were 95% reliable, the overall reliability of the entire assembly would then be

$$95\% \times 95\% \times 95\% \times 95\% \times 95\%,$$

or 77% at best, for in addition to the unreliability factor of each stage is the hazard of disengaging each stage from its booster.

The second drawback of staging is the monstrous weight which we must place on the launching pad for each kilogram of ultimate payload. To see this clearly, suppose the weight of each rocket stage were ten times the weight of its payload, containing, remember, the entire next rocket stage. For the five-stage rocket, the ratio of the final payload weight to that of the fully loaded rocket is

$$\frac{1}{10} \times \frac{1}{10} \times \frac{1}{10} \times \frac{1}{10} \times \frac{1}{10},$$

a hundred thousand kilograms on the pad for each kilogram of payload!

The *space shuttle* is a sensible concept for avoiding the high cost of space voyages, that is, for avoiding the high cost of jettisoning used-up rockets during each journey. The shuttle is a reusable space vehicle which also can take advantage of the atmosphere in obtaining aerodynamic thrust and lift.

It is natural to wonder which of the following two alternatives is preferred for bringing a payload to its greatest distance against an opposing gravitational field: fire an upper stage after its booster has carried it to its maximum height, or fire it instead at a low altitude immediately following the burnout of its booster's propellant.

The answer can be found very quickly: Let us assume that we have a two-stage rocket and each stage, of unit mass, has the same burnout speed. In the first case, the booster carries the upper stage to a maximum height h_0 found from the expression in which all the kinetic energy of motion is transferred into potential energy of height.* The upper stage is then fired from rest and its payload is carried up to a height h_0 from that point or $2h_0$ from the launching pad.

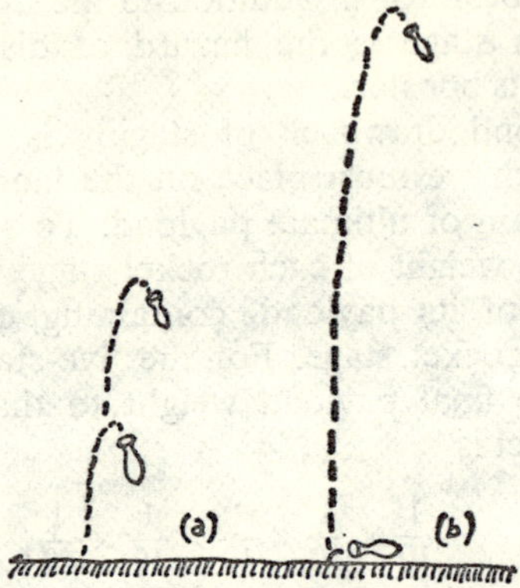

FIG. 25. The upper stage of the rocket in (a) is fired when its booster's *apogee* is reached. Assuming that the booster and upper stage are similar in that their "burn-out" speeds are the same, each will reach the same apogee height from launch as shown. The rocket in (b) is exactly the same as in (a); its thrust program, however, is different: The upper stage here is fired close to the ground, immediately following its booster's propellant burnout. Now the apogee of the upper stage is actually *twice* as great as the one in (a). We assume in (b) that the force of gravity is constant at all heights, air drag is negligible, and that propellant burnout occurs almost immediately.

* The energy equation for either stage of Case One is $\frac{1}{2}v_0^2 = g_0 h_0$, where v_0 is the burnout speed. It has been assumed for simplicity that the thrust period is short and that g_0, the gravity force field at the Earth's surface, is unchanged all along the vertical rise up to h_0 (this is the "flat Earth" approximation which holds fairly well so long as h_0 is a very small fraction of the Earth's radius).

In the second case, the upper stage is fired at twice the speed near the launching pad so it has *four times* the kinetic energy as in Case One. This means that the payload in Case Two reaches *twice* the height of that of Case One!†

It may seem paradoxical to be able to get the greatest benefit from a propellant not by firing a given stage when it is as far out as possible, but rather at the moment when it is moving out *as fast as possible* which may occur, in fact, at its minimum altitude.

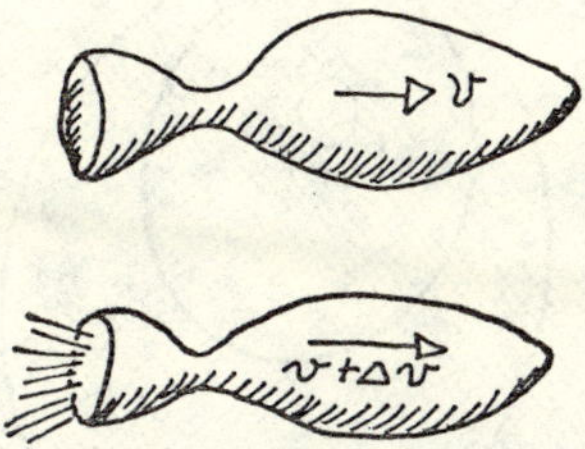

FIG. 26. If an object, moving along at some speed v, gets a boost in the same direction by an amount Δv, then its increase in kinetic energy is $v\Delta v + \frac{1}{2}(\Delta v)^2$. Thus, if a rocket receives a small boost in speed, say from turning on its engine for a short time, this can give the rocket a very large *increase* in kinetic energy *provided it was already moving fast before the boost.*

Anything moving at a given speed v, which gets an additional boost in speed Δv will move at a final speed

$$v + \Delta v ,$$

† The upper stage of Case Two reaches a height h_1 from the launching pad as found from:

$$\tfrac{1}{2}(2v_0)^2 = g_0 h_1 .$$

By comparing this equation with the similar one above for Case One, we can see that

$$h_1 = 4h_0 .$$

of course. But the *kinetic energy per unit mass it has gained,* which, for example, can be used to overcome gravity, could be large. For example, if the boost in speed that is given a moving object is fixed, as the boost given a rocket by its engine firing for a fixed time in a given gravity field, then the rocket will receive a tremendous boost in energy *when just prior to the boost it is moving rapidly.*‡ But when *is* a rocket moving rapidly? Two cases are obvious–the instant after it has been boosted by a previous booster stage, or, if it is

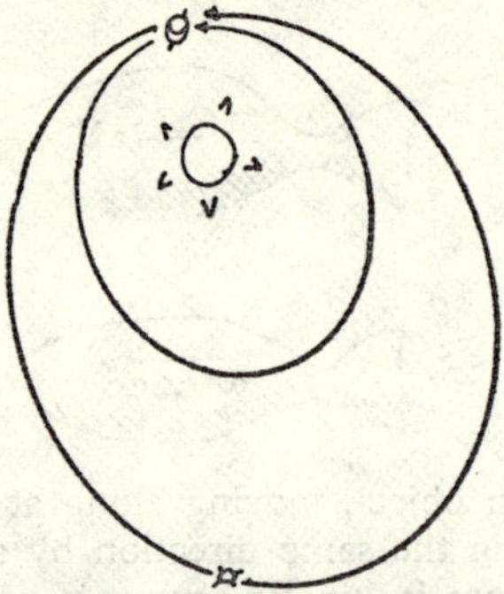

FIG. 27. The first Soviet "planetoid" was launched into an escape orbit relative to the Earth and in the direction of the Earth's orbital motion, on 2 January 1959. By choosing the Earth's perihelion day for the launching, the rocket was given the maximum possible boost by the Earth's orbital velocity which would enable it to reach as far out into the Solar System as it could. This would happen even though the launching took place while the Earth was as close to the Sun as it could get! In this example the v of Figure 26 is the Earth's perihelion speed and Δv is the speed of the rocket relative to Earth after escaping Earth's field.

‡ The kinetic energy gained is

$$\tfrac{1}{2}(v+\Delta v)^2 - \tfrac{1}{2}v^2.$$

We divide this by Δv to get the energy gain per given boost in speed,

$$v + \tfrac{1}{2}\Delta v.$$

in an elliptic orbit, the instant it is closest to the gravitational center, at *perigee* in the case of an Earth orbit.

In the early years of the Space Age both the Soviet Union and United States were attempting "space spectaculars." The Soviet Union had a clear edge. In 1959 they attempted to launch a planetoid out of the Earth's orbit into the Solar System, and succeeded. The launching date was January 2nd. This was the very date on which the Earth was at *perihelion,* its closest distance to the Sun. They were obviously making use of the principle above—launching the spacecraft in the direction of the Earth's motion in the Solar System when the Earth is moving most rapidly in the Solar System. What is surprising is that the rocket was able to move furthest out in the Solar System by selecting the date of launch to be the one when the Earth and it were furthest in.

The most interesting example of this paradox occurs when a rocket spacecraft in circular orbit wishes to move far out of its gravitational field and still have kinetic energy left over. For example, suppose an Earth satellite wishes to move far out in the Solar System toward another planet. Obviously it must rid itself of the Earth's gravity yet still have orbital energy left to carry itself to, say, Mars or Jupiter. The question to be asked is this: Could it be actually worthwhile to *cancel* some of the hard-gained circular orbital energy about the Earth by a reverse thrust and then fall in close to the Earth, moving therefore at high speed at perigee and *there* blast off for the distant planet? The answer is yes, under a special condition: If you wish not only to escape a gravity field completely, but have a large amount of kinetic energy left over after you do. How much energy? At least more than the minimum energy needed to escape from a stationary point in the original circular orbit. Under this condition it would actually pay to cancel some of the circular velocity, fall in

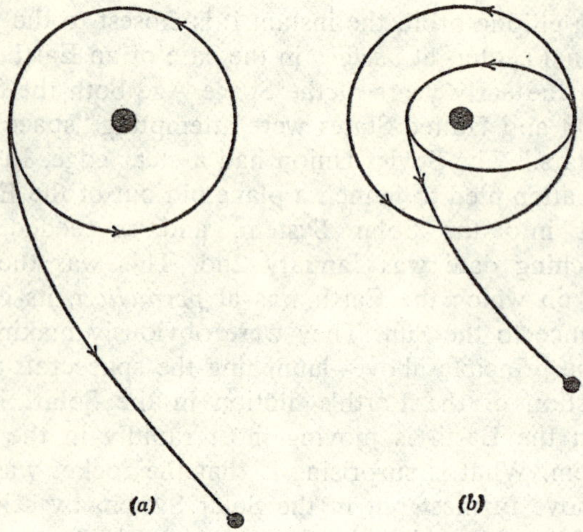

FIG. 28. Alternative ways to escape from a circular orbit to "infinity": In (a) the rocket simply blasts off into a parabola or hyperbola. In (b) the rocket *reverses thrust* to fall into a close-in ellipse and then at the *pericenter* or *near apsis* of the ellipse (perigee if orbiting about the Earth) it blasts off into a parabola or hyperbola. Despite the loss of circular orbital energy in maneuver (b), it is preferred to (a) provided the energy remaining in the hyperbola after completely leaving the gravitational field exceeds the escape speed at the radius of the circular orbit. (If the arrows were reversed the figures would represent alternative maneuvers for *entering* a circular orbit from infinity.)

close, and then blast off "to infinity." This would be the state of affairs which would occur if a space station in a high orbit were to send space vehicles to many different destinations. But in planning an interplanetary trip in advance from a launchpad on Earth, one would scarcely attempt to place the spacecraft in a *large* circle about the Earth as an initial maneuver.

Trajectory of an ion rocket

Let us assume that a two- or three-stage rocket has managed to inject its payload—a rocket which can produce a very small measured thrust—into a low circular orbit, yet sufficiently far out in space so that air drag no longer exists. We thrust horizontally, that is, tangent to the circle, and control the thrust so that it is as small

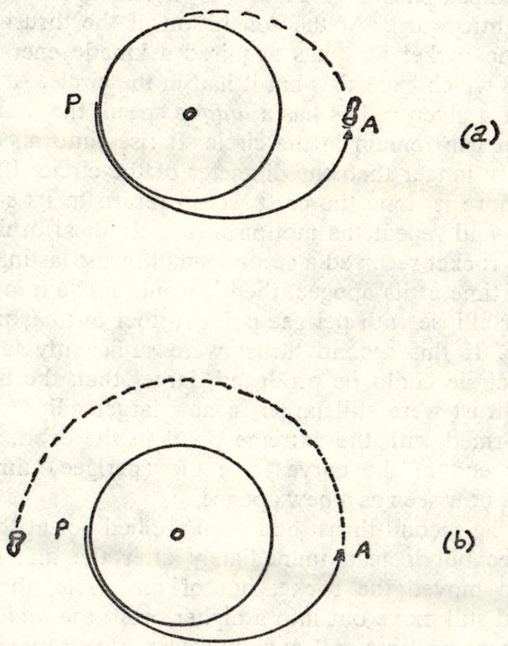

FIG. 29. (a) A rocket-satellite is given a thrust for a short time at *P*. This will bring it to *A*.

(b) It is given a second thrust at *A*. This will make the new ellipse formed more circular and the orbit larger.

as we wish yet still enough to give us a measured result in terms of changed speed and direction of flight. As we discussed previously, unlike a rocket standing on its launching pad, anything in orbit is freely falling, without any support or weightless, and the smallest thrust can alter the trajectory.

If this applied thrust is small compared with the force of gravity, and if it lasts for a time that is small compared with the orbital period, then the orbit will be the ellipse shown in Figure 29(a). The reasoning is straightforward: At the conclusion of the thrust interval the rocket now has acquired a kinetic energy and speed which exceeds what it had in the circle. As a circle of a given radius has a *unique* speed, the rocket no longer can remain in the circle. It rises into an ellipse slightly longer than the diameter of the circle. If nothing more is done the craft would return to its starting point and repeat the motion in the elliptical orbit. But if the rocket received a second small thrust lasting for a short time at its apogee, then it would move into a still larger ellipse with perigee point further out beyond the circle. If this second thrust were sufficiently large, a new circle could be produced, larger than the first. If the thrust were still larger, a new larger ellipse would be formed with the extreme point in the orbit, at the other end of the curve from the (perigee) thrusting point, now seen as a new apogee.

If the second thrust had been applied not at the first apogee but instead immediately after the first thrust, which moved the rocket out of the circle, the craft would still move out into a higher orbit; the *major axis* of the new ellipse will now, however, have *turned* compared with the old. (The major axis is the line joining the two extreme points of an ellipse.)

If we were to give our rocket, originally orbiting in a circle, a series of *short, frequent microthrusts,* the trajectory would be made up of short arcs of a family of ellipses. If the time interval between thrusts were negligible compared with the period of the ellipse which would be formed by ceasing the thrusts, the *powered trajectory* then produced would be a *spiral*.

Thus, an ion rocket which takes off from a circular orbit with a continuous thrust would move about the Earth in an ever-widening arc. This path could resemble the groove of a phonograph record if the thrust were very small compared with the local force of gravity. If the thrust were larger the pitch of the spiral would be larger. In order that a so-called *logarithmic spiral* be produced, one in which the angle between the trajectory and the tangent to a circular orbit remains constant, as in the groove of a phonograph record, then the thrust force divided by the local gravity force also must stay constant. This means that the thrust force would weaken as the rocket moves further out from the gravitational center at the same rate as gravity weakens, that is, by the $1/r^2$ or inverse-square law. It is far more likely that we would want an ion rocket to maintain its thrust constant at its rated value, since this would enable it to move out to high altitudes sooner, in ever-widening arcs.

The satellite energy paradox

In Chapter Two we stated that the work required to move a one kilogram mass from the Earth's surface against the opposing gravity force field to some height h that is small compared with the Earth's radius (the flat Earth approximation) is gh—the greater the height the greater the work needed to oppose gravity, in direct proportion to h. The gravity force field's *capability* of doing work at height h is called its *potential* at h, for if we released the kilogram at h, the kinetic energy produced by the force field on striking the Earth is just gh.

In this simple analysis g is considered to be constant,

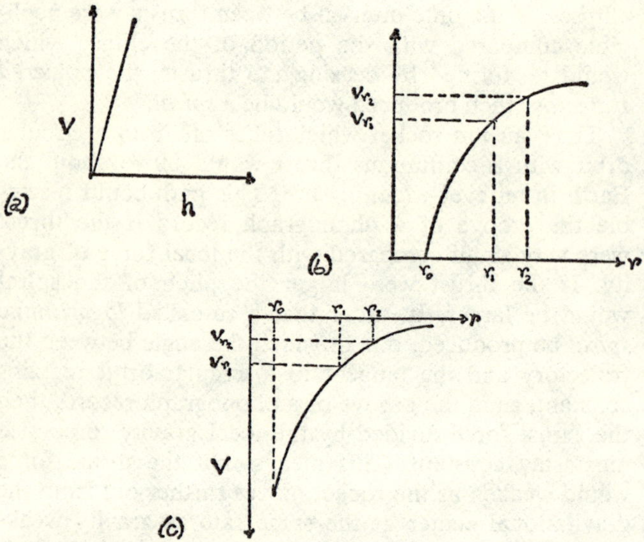

FIG. 30. (a) If the Earth were flat, the *potential* (that is, potential energy per kilogram) measured from the Earth's surface would increase linearly with height. This means that if you dropped a rock from rest at 20 meters it would hit the ground with twice the energy than it would if you dropped it from 10 meters.

(b) In reality the Earth is almost a sphere; therefore it is more correct to show the potential curve this way. This is the potential, V, as a function of distance from the Earth's center although the curve starts at r_0, the Earth's radius. Notice that the beginning of the curve is almost a straight line.

(c) We are really interested only in *differences of potential* in moving from one point in space to another, so it really doesn't matter where we place the zero potential level. Here V approaches zero as r approaches infinity. But the same work is needed to move an object from r_1 to r_2 as was needed in (b). By choosing the zero reference level at infinity, all values of potential are negative, but some are more negative than others (those at points closest to the Earth).

equal to the value at the Earth's surface which we call g_0. But when the variation of g with height was considered to be important, that is, when the range of study is comparable to the Earth's radius, the potential calculation became more complicated. When the Earth is considered as a sphere, g above the surface diminishes with the square of the distance from the Earth's center. We then must calculate the so-called *line integral* of a (variable) force through a distance—the concept of work. The result of the calculation predicts, of course, that the further we move outward from the Earth the larger will be the potential. But at great distances the gravity force has little effect on the calculation and so the potential curve bends over.

The result of the calculation of potential has a very simple form if we shift the zero reference level from the Earth's surface all the way out to a place, "infinity," where the gravity force field is itself zero. This, of course, will have the highest possible value of potential, and the potential at every other place will therefore be *negative*. As potential has meaning only in terms of its value at one point in space *relative* to its value at another, then it does not really matter where we set the zero level. When we work through the calculation we get the simple result:

$$\text{potential at } r = -\frac{\text{constant}}{r}.$$

We can call the constant "k", but in the calculation it turns out to be $g_0 r_0^2$ where r_0 is the radius of the sphere.

Now we are led to the question: Given a satellite in a circular orbit of radius r, what is the relation between its kinetic energy per kilogram, or *kinetic*, to its potential energy per kilogram, or *potential?*

It turns out that the *magnitude* of the potential is twice as great as the kinetic. As the potential is actually negative, the potential–kinetic sum is also negative and

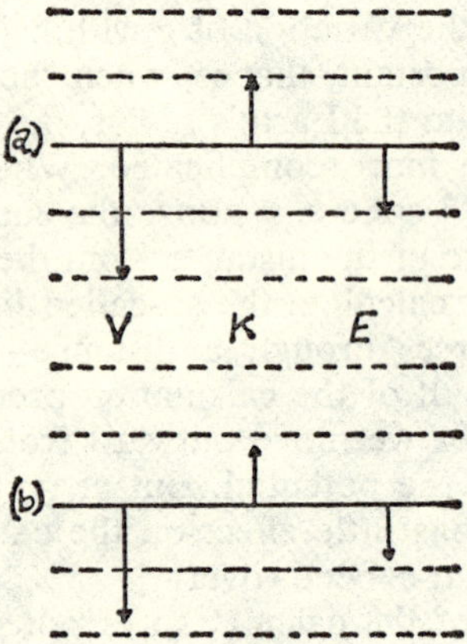

FIG. 31. (a) Potential (V), kinetic (K), and their sum, total energy (E) for a satellite in a circular orbit.

In (b) the satellite has received a very small thrust (compared with the force of gravity) throughout the circular orbit. In receiving this added energy the orbit has now opened into a *spiral*, a "quasicircular" path in which V, K, and E continue to bear the same relationship to each other as in (a).

also half as large as the potential.* The three energies are shown relative to each other in Figure 31.

Suppose a microthrust were given to a rocket in circular orbit, yet so weak as to barely disturb the orbital motion. The thrust is applied along the direction of motion and lasts at least for one orbital period. The

* For any object moving at speed v_c, its kinetic is $K = ½v_c^2$. But we found in Chapter Two that $v_c^2 = gr$, and $g = \frac{k}{r^2}$, so that the kinetic: $K = \frac{k}{2r}$. Since $V = -\frac{k}{r}$, the *total energy*, E $(= V + K)$ is therefore $E = -\frac{k}{2r}$.

rocket's path resembles a "phonograph record groove," which we can call a *quasicircular* trajectory, for the gravity field so strongly dominates the dynamics that the craft could be said to be almost in a circular orbit. The word *orbit* refers to a closed curve, so we reserve the word *trajectory* or *path* for powered flight; we will not use the term orbit for parabolic or hyperbolic free-fall flight paths, although many people do.

The microthrust, applied to the path length, adds energy to the rocket. The total energy, *E,* must therefore *increase,* that is, *become less negative.* This concept of negative total energy might be confusing unless we recall that the determining factor is the large negative potential energy based upon the *arbitrary* setting of its zero level at infinity. The potential, in this quasicircular orbital state, is two times the total energy, and so the potential must also increase as the rocket moves further out beyond the original circle. *To maintain the energy relationships in a circular orbit,* it follows that *the kinetic energy will decrease.* For each unit of energy gain, resulting from the microthrusts, the potential gains two units and the kinetic loses one.

In the reverse situation, a satellite in circular orbit encounters a small amount of atmospheric drag. This opposing force results in a loss of energy. For each unit of energy lost, the satellite will lose two units of potential as it spirals down, but will *gain* one unit of kinetic.

Although the satellite encountering an opposing drag force will *speed up* as it slowly spirals in, this situation will not continue indefinitely. Gradually the drag force becomes stronger and stronger while the gravity increase is relatively slight until drag is no longer a small perturbative force but completely dominates the dynamical problem. When that happens air drag acts as a true braking force and *slows down* the falling satellite.

One can draw a parallel between the trajectory of a satellite under an increasing drag-to-gravity ratio and that of a rocket orbiting in a circle under an increasing thrust-to-gravity ratio. Assume first that the thrust is small enough to maintain the quasicircular condition, so that it moves more and more slowly as it spirals around the gravitational center. Eventually the rocket will have gone so far out that the thrust force, rather than gravity, dominates the dynamics even though the thrust is maintained constant at its small value. And when that happens the rocket moves in a wide arc and *speeds up* as a result of the applied thrust.

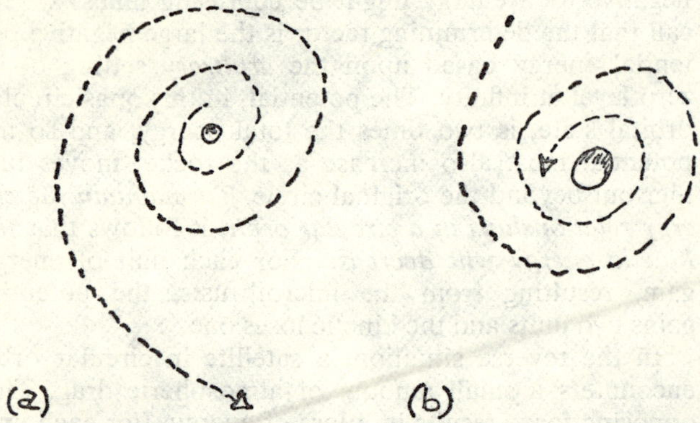

FIG. 32. (a) A small thrust applied continuously pulls a satellite out of a circular orbit into a spiral. This will continuously *slow down* the rocket-satellite as it moves further and further out. Eventually gravity becomes so weak that the thrust force plays the principal role in controlling the path; the thrust will now speed up the rocket.

(b) A satellite encountering atmospheric drag. Initially, the drag force, which takes away orbital energy, is so weak compared with gravity that as the satellite begins to spiral in because of the drag it *speeds up* as it falls. However, as the drag force increases in the lower atmosphere the drag eventually acts as a brake on the motion *slowing down* the satellite as it plunges to Earth.

One of the more difficult space dynamics problems to analyze is one which looks as if it ought to be simple: a rocket moves in a gravity field and also under the influence of a thrust of constant magnitude aligned straight along the instantaneous direction of motion of the rocket. The difficulties are compounded by the depletion of rocket mass as the thrust continues in time, for even though the thrust is constant, the acceleration due to thrust will continue to increase as propellant is depleted. In the long run we must analyze the problem with the aid of a computer. By "analyze" we mean finding the velocity and position of the rocket at future and past times from the known variation of thrust and rocket mass with time, the way the gravity field varies from point to point in space, and the rocket's velocity and position at one specific time.

The Moon as ion rocket

In the last century a curious discovery was made: astronomical events in the Solar System were simply not running on schedule. At least it appeared that way. What caused it? Was time really running more slowly? Or perhaps merely our measurement of it was at fault? The rotation of the Earth, tied ultimately to the position of stars on the cross hairs of telescopes, was the primary standard of time. And herein lay the difficulty. Only in recent years, when the ultraprecision of atomic clocks, tuned to the frequency of vibrating atoms, was matched against the Earth's rotational period, could a precise figure be set down for what turned out to be the lengthening of each day. Each day is but 0.0018 seconds longer than one a hundred years ago, but considering that there are almost 37,000 days in a century, the cumulative error can be rather disturbing.

The cause of this error is not difficult to understand. If the Earth were a rigid body left to rotate alone in space, its continuous spinning would be the closest we

FIG. 33. The "tidal bulge" of the Earth toward the Moon and Sun.

could come to a perpetual motion machine. But it is neither rigid nor is it alone. Even forgetting for the moment the influence of its partner in space, the Moon, portions of the Earth are closer to the Sun than its center while other portions are further away. The unequal force of attraction tends to elongate the Earth along its direction to the Sun. Because of the Earth's rotation, the elongation at any given location is very temporary. Large masses of the Earth's bulk are therefore stretched and strained periodically. Tremendous volumes of water are bulked to and fro across the Earth's surface in an attempt to follow the variable gravity field of the Sun in which the Earth is turning. The energy dissipation of Earth's rotational energy by this tidal friction is enormous.

The effect is even more pronounced when one considers the mutual attraction of the Earth and Moon. Because of the Moon's diminutive mass compared with the Sun's, it might appear that the Moon–Earth tidal effect could not possibly be so important. But our concern is with a *differential* gravity effect–the *change* in the Moon's gravity as measured across the volume of the Earth. Gravity, an inverse-square force, is extremely sensitive to the distance from its source. Thus, although the Moon's attraction to the Earth is rather small compared with the Sun's, there is a relatively high percent difference in the Moon's attraction of a near portion of the Earth compared with that of a distant portion—an *inverse-cube* effect. Therefore the Moon, far more than the Sun, continuously stresses the Earth and disrupts what would otherwise have been a precision rotation.

But consider the effect on the Moon's motion: In 24 hours the Earth turns completely around, but the Moon turns a bare 13° about the Earth in the same direction. The oceans bulk toward the Moon, and in doing so resist the relatively rapid rotation of the solid Earth beneath them. This elongated mass of water is pushed slightly ahead. The Moon is aware of this angular displacement, for it finds its attraction to the Earth not at Earth's center but displaced off-center. A component of the gravitational force lies permanently in the direction of the Moon's orbital motion!

Is the Moon, therefore, in a closed orbit about the Earth, or is it more like a powered ion rocket spiraling further and further away? I will leave it to the reader to decide. But where does the power come from? It is actually transferred out of some of the lost kinetic energy of the Earth's rotation directly to the Moon. Ignoring the effect of the Sun, the Earth–Moon angular momentum is maintained constant; the Earth's angular momentum drops as the Moon's angular momentum rises.

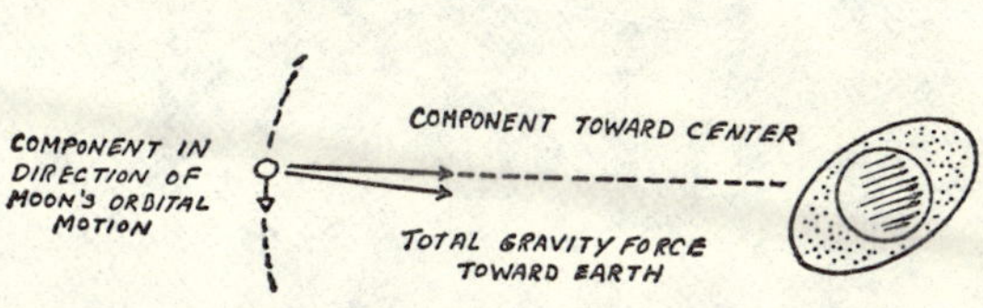

FIG. 34. The Moon as "ion rocket." The tidal bulge is pushed slightly ahead by the rapidly rotating Earth; therefore, the gravitational attraction of the Moon has a slight force component *in the direction of its orbital motion.* This accounts for a barely perceptible outward spiral path.

Astronomers have looked into the future and have set a date when both the day and month will have the same duration, about 60 of our present days. At that time the outward spiraling will cease, but the Sun's tidal effect will then become dominant over that of the Moon, causing the tidal bulge to lag *behind* the Moon's motion. This will act as a drag force on the Moon, and like any satellite encountering drag, it will spiral in toward the Earth. The final cataclysm lies billions of years into the future, but one question is still unanswered: Will the oceans still be around?

CHAPTER FIVE

Sailboats and Slow Boats to Phobos

Gulliver

"Certain astronomers . . . have likewise discovered two . . . satellites, which revolve about Mars, whereof the innermost is distant from the center of the primary planet exactly three of its diameters, and the outermost five; the former revolves in the space of ten hours, and the latter in twenty-one and a half, so that the squares of their periodical times are very near in the same proportion with the cubes of their distance from the center of Mars, which evidently shows them to be governed by the same law of gravitation that influences the other heavenly bodies."

Thus did Jonathan Swift, in his sparkling satire of 1726, *Gulliver's Travels,* relate the astonishing achievement of the astronomers of Laputa, the island in the sky. The achievement was indeed Swift's, for the two satellites of Mars—Phobos and Deimos, mythical sons of the war god, Mars, named after Fear and Terror—were not really discovered until 1877! They can be observed only through very large telescopes which were unknown in Swift's time.

Phobos, the innermost satellite, actually revolves in just under eight hours. Although it moves about the planet in the conventional direction, from west to east, it is unconventional in that it revolves in one-third the time that the planet rotates, and so, unique among all natural satellites in the Solar System, it rises in Mars's western sky and sets in the east. Deimos, the second satellite, revolves in slightly more than thirty hours. How Swift "knew" about them remains something of a mystery, although it must be admitted that more than a century earlier, Kepler "deduced" that Mars must have two satellites because Earth has one and Jupiter "has four." The law of geometric progression assigns the number two to the planet in between.

If Swift had written his book in the sixteenth century, it would have been considered to be less a mystery than a heresy. Few writers would have dared to describe a model in the sky of satellites that circulate about a large central body, much as the planets move about the Sun.

Although future historians may regard the stepping off from our own planet as a turning point of modern history, there is little doubt that the discovery of space flight's physical principles marked another turning point—the end of medieval times. From Copernicus to Kepler to Galileo to Newton, the Aristotelian myth—a man-oriented Universe—yielded to the radical new concept—a mechanically oriented Universe. It ran like a gigantic clock, fully detached from humankind.

"There are then innumerable Suns, and an infinite number of Earths revolve around those Suns, just as the seven we can observe revolve around this Sun which is close to us." For such a statement was Giordano Bruno burned at the stake in 1600. Excessive punishment is the mark of a desperate society. In that same year Tycho de Brahe and Johannes Kepler, architects of the new Universe, met for the first time. The world would never again be the same. By the close of that century Isaac Newton had united the new celestial physics of Kepler with the terrestrial physics of Galileo in a synthesis—the universal law of gravity and the dynamical laws of motion—that drove the final nails into the coffin of the ancient world.

The three laws of Kepler

Events and discoveries have a way of building one upon the other to erect an edifice, or bring another down, or both. To a very crude approximation the orbits of the planets were indeed the circles of Copernicus. Even he succumbed to the authoritarian bias of the ancient world that held up the circle as the perfect form. For the planets no other arc was possible.

Of the planets known in Kepler's time only Mercury's orbit departed noticeably from a circle. But Mercury followed the Sun closely in the heavens; it could barely be seen much less studied. Of all the other planets, Mars, the bright red spot in the night sky, proved to depart most from a circle, though by a mere half percent. Only the refined observations of Tycho, good to within two or three minutes of arc, could have detected such a small departure. Pre-Tychonic astronomers could do no better than ten arc minutes in precision or about one-third of the Moon's diameter as seen from the Earth. What a stroke of luck that Kepler's search for the solution of the Solar System mystery should have coincided with Tycho's deliverance of positional data on nothing less than Mars itself!

But what if the telescope, invented less than a decade later, were applied to the problem? Would Kepler's discovery of the pure elliptical orbit have been thwarted by perturbative effects? How would Kepler have coped with data that would have proved to be *too* accurate, destroying his elegant ellipses by the yet-to-be discovered gravitational attraction of the planets upon each other? Would he have been able to sort out these separate influences and make the intuitive leap to the inverse-square gravitational theory in terms of the force between planet and Sun, planet and satellite, and planet and planet? This task remained for Isaac Newton; standing on the shoulders of Kepler and Galileo he saw further and deeper.

Almost out of nowhere, at just the right moment, the telescope appeared. As the story goes, a man at the Frankfurt annual fair of 1608 offered one for sale; the following month Johann Lippershey, a spectacle maker, applied in the Netherlands for an exclusive license for its manufacture. The application was denied because of conflicting claims to the invention. In the following summer Galileo constructed his own instrument after hearing reports of the Dutch invention, and became the first to turn the telescope to the sky and report his findings.

The telescope revealed tens of thousands of *individual* stars in an almost limitless Milky Way sea. It immediately raised the theological question: Could the firmament really exist only for man's pleasure when it could not even be truly seen except with this instrument? The Universe no longer was a tight-fitting hemisphere, but expansive beyond the dreams of humankind. Its boundless horizons were soon to be described by John Milton in *Paradise Lost:*

> A broad and ample road
>   whose dust is gold,
> And pavement stars, as stars
>   to thee appear
> Seen in the Galaxy, that Milky Way
> Which nightly as a circling zone
>   thou seest
> Powder'd with stars.

When he trained his glass on Jupiter, Galileo found four moons, a Copernican Solar System in miniature. So enraged were the classicists of his time that they refused to gaze through the "devilish tube" and see "the wonders" with their own eyes. His report on the findings, *The Star Messenger,* spread like wildfire across Europe. And so did his enemies. Only one voice came to his defense, the mighty voice of Johannes Kepler.

Kepler believed that the motion of the single planet Mars held the secret of *all* planetary motion, a perfect example of inductive reasoning–from the specific case to the general principle. This "assault on Mars" was initially planned as a few months' work but turned out to be more than six years of painstaking labor. In contrast to *The Star Messenger,* read and understood in a few hours, Kepler's monumental *New Astronomy,* some have complained, would take a lifetime to comprehend.

Without immediately abandoning the circles of Copernicus, Kepler did clear away a basic Copernican obstruction to progress: To maintain the ancient bias of uniform circular motion and therefore not produce a theory at variance with known observations, Copernicus presumed that the common center of all planetary orbits was somewhat displaced from the Sun. Kepler felt, however, that a shift of the center from the Sun and the concept of uniform circular motion were absurdities, firstly because the essential "force" must surely arise from the Sun, and secondly because uniform motion was a bias without grounds.

I will avoid recounting Kepler's years of struggle with Mars, in which he abandoned the circle for the oval and the oval for the ellipse. His final conclusion was his *first law* of planetary motion: *The orbit of each planet is an ellipse with the Sun at one focus.*

What is an ellipse? For one thing, it is a *plane* figure as is the circle. For another, similar to the circle, it is a closed curve: An object tracing out an ellipse will return to its starting point. But to define an ellipse and see how it differs from a circle, let us set down two points and require that we are allowed to move anywhere in a plane as long as the *sum* of the distances to each of these two points is always the same. This idea is most easily demonstrated by placing a pencil at the

(a)

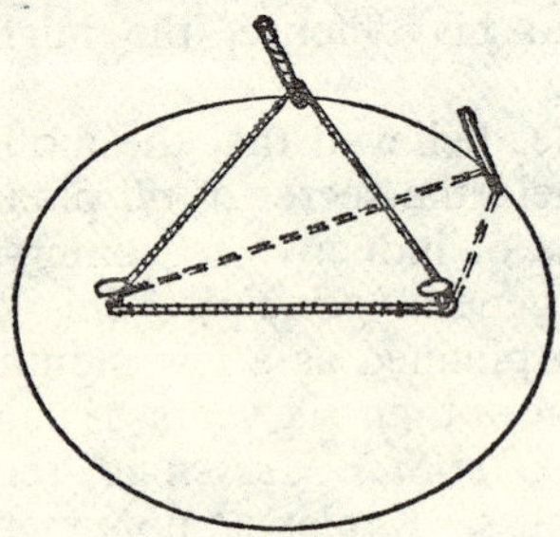

(b)

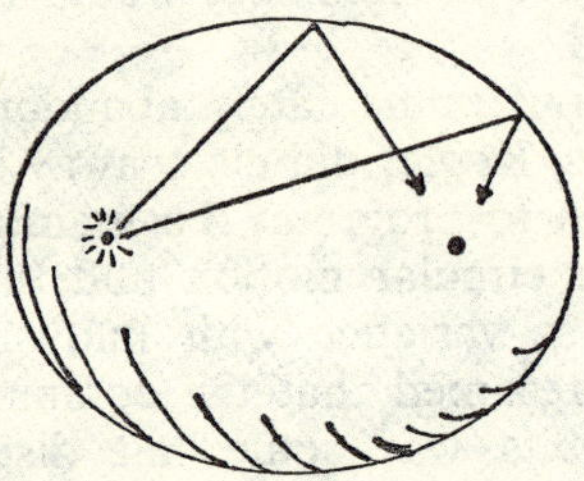

FIG. 35. (a) An ellipse is *produced* with two nails, a string, and a pencil. It is *defined* as a curve formed by every point in a plane in which the sum of the distances to two fixed points (called *foci*) is a constant. It is not hard to see that this sum is equal to the length, or *major axis,* of the ellipse.

(b) An *ellipsoid* is the solid figure produced by rotating an ellipse about its major axis. Rays from one focus reflected from the smooth surface of an ellipsoid obey the law of reflection (equal angles are made with the normal to the surface) and will move straight toward the other focus.

end of a triangle formed by a string tied to each of these points, as shown in Figure 35, and then moving the pencil about so that the string always remains taut. If the point of the pencil were not constrained to move in a plane then the three-dimensional figure it would trace out would be an *ellipsoid*. This is the figure which would be produced by rotating an ellipse about an axis

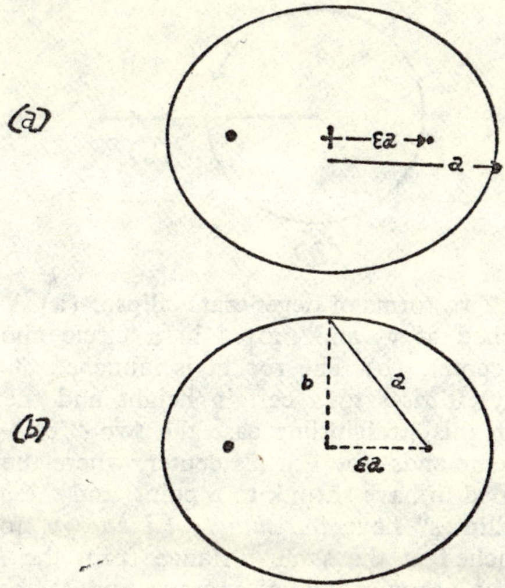

FIG. 36. (a) The eccentricity, ϵ, tells us how far the focus is from the center in units of a.

(b) In general $c + d = 2a$, where c and d are the distances to the foci from a point on the curve. In the special case where the point on the ellipse is at the end of the *semiminor* axis b, each line to the focus is of length a.*

through the original two points, or *foci*, of the ellipse or ellipsoid. One well-known property of ellipsoids is that if you place a source of light, or sound, at either of the two foci the rays will reflect from the ellipsoidal surface and converge right to the other focus. Some ceilings have been built in this shape in cathedrals, and a whisper near one end can be heard near the other at the distant focus.

It is obvious enough that if the two foci merged into one the ellipse would turn into a circle. But even if the

* By applying the *Pythagorean theorem* to the right triangle thus formed, we find that $b/a = \sqrt{1 - \epsilon^2}$. We can then see what will happen to b if ϵ is close to zero or to one which are its two possible extreme values.

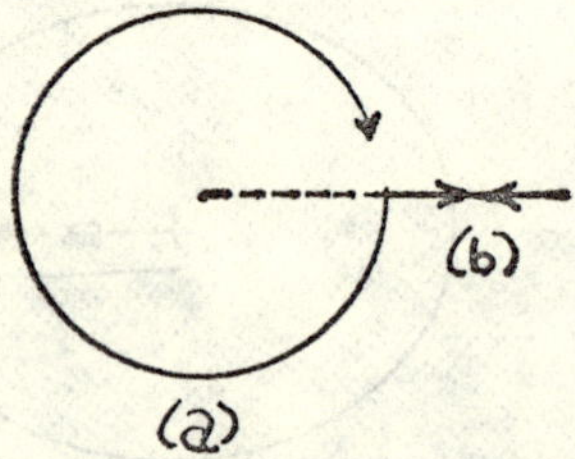

FIG. 37. Two forms of degenerate ellipse: (a) A rocket is launched at v_c and moves in a circle about the Earth's center. (b) The rocket is launched at v_c but vertically. It rises to a certain height and then falls down. In this straight-line case the two "foci" are at the extreme ends—the Earth's center, where the Earth is imagined to have shrunk to a point, and the apogee. Both "ellipses" have the *same total energy,* for they are launched at the same distance from the Earth's center (i.e., same potential energy) and at the same speed (i.e., same kinetic energy). The *size* of an orbit—its *a*—depends on its total energy (per unit orbiting mass) *alone.* Hence each orbit will have the same length; therefore the vertical rise will be exactly equal to the Earth's radius.

foci were separated the figure formed would still almost be circular, provided the distance from the curve to either focus were much larger than the spacing between the foci. It is easy to see, by choosing the extreme case in any ellipse when the pencil is in line with the foci, that the average distance from one focus to the pencil, or half the sum from the two foci, is the half-length of the ellipse. The full-length is the *major axis,* so the average distance from a focus to the ellipse is equal to the *semimajor axis.* The departure of an ellipse from a circle is most easily described in terms of the location of a focus from the center of the ellipse as some fraction of the distance to the end, i.e., by stating the distance of a focus from the center as a certain fraction of the semi-

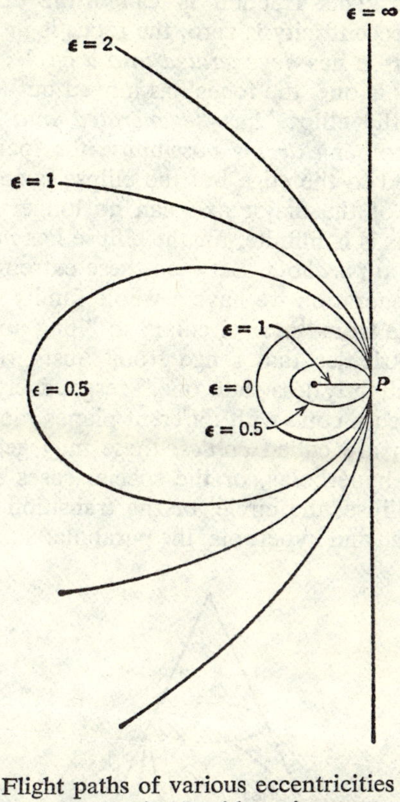

FIG. 38. Flight paths of various eccentricities resulting from launchings at point *P* with various energies. (Eccentricities greater than 1 refer to hyperbolas and again express the distance of a focus from the origin. Hyperbolas occur in pairs and the origin lies midway between their closest points, which are joined by the major axis.) The black dot is at a focus, the source of gravitational attraction for all these paths. An eccentricity of one refers either to a parabola (an "ellipse" with infinite major axis), or to a straight line (an "ellipse" with infinitesimal minor axis) as in Figure 37.

major axis. This fraction is called the *eccentricity.* When the eccentricity is zero, the focus is *at* the center and the ellipse has *degenerated into a circle.* When the eccentricity is one, the focus has moved out to the very edge and the ellipse has *degenerated into a straight line.* As another *extreme* possibility, the focus has not quite moved to the edge but the ellipse is so long that the length of the major axis can no longer be found; the major axis is infinite, for the ellipse has now *degenerated into a parabola.* Between these extreme cases of ellipse degeneration we have a whole family of ellipses which range from "near circular" to "long and skinny" with eccentricities that range from "just greater than zero" to "slightly less than one," respectively.

If we slice a cone with different planes the curves of intersection are called *conics;* these may generally be ellipses or hyperbolas, or the special cases of the degenerate ellipse, the circle, or the transition curve between ellipse and hyperbola, the parabola.

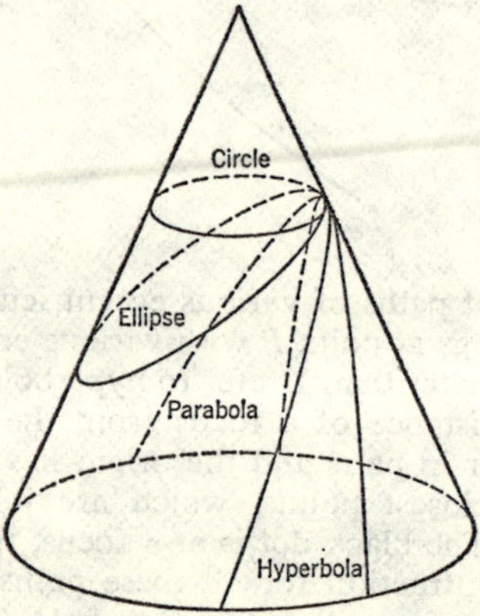

FIG. 39. The intersections of planes with a cone produce the curves of orbital space flight about a single gravitational center.

All of these curves are of interest to us in space dynamics. Comets move about the Sun generally in long ellipses of great eccentricity, or possibly in hyperbolas (or again the transition case, the parabolas). In elliptical paths the comets reappear periodically, as for example Halley's comet, but in hyperbolas they are seen once but never again. Comets of this second class either enter the Solar System with some energy of motion relative to the Sun, or else they somehow acquire additional energy in a close encounter with a massive orbiting object such as Jupiter.

Space vehicles, such as Mariner and Viking, which

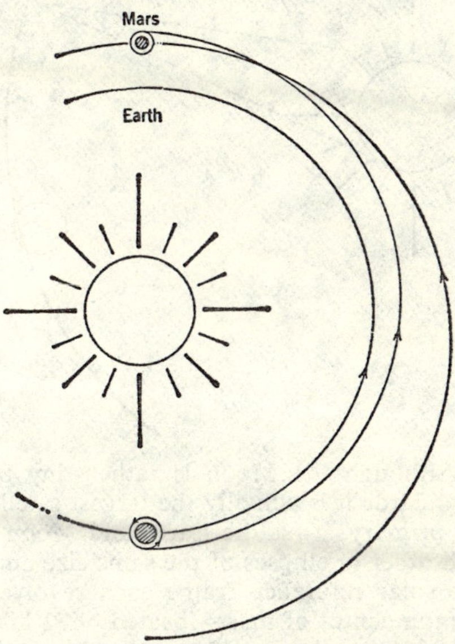

FIG. 40. A circular satellite orbit about Earth is cotangent with a hyperbola about Earth which then merges with an interplanetary ellipse about the Sun. The size of the planets is, of course, *very* much smaller than their orbits about the Sun, or the Sun itself.

travel to other planets, must depart in a hyperbola *relative to the Earth* which merges with an ellipse *relative to the Sun,* having end points at the orbits of the Earth and the destination planet.

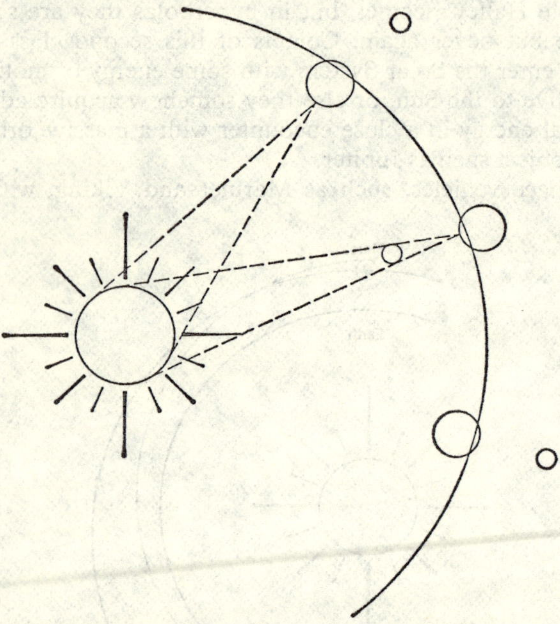

FIG. 41. Although the Moon is rather tiny by Solar System standards it is actually the largest satellite relative to its primary. Thus the Earth and Moon revolve about each other in ellipses of the same size and shape, but in the solar reference frame each revolves about their common center of mass–located 4672 kilometers (2903 miles) from the Earth's center, that is, 73% of the distance from center to edge–and the "double planet" moves about the Sun in an ellipse. The Earth–Moon center of mass follows this ellipse while the Earth's center oscillates back and forth about this curve and the Moon's orbit is always concave toward the Sun.

To make *Kepler's first law* as useful as possible we might generalize it this way:

> *The paths of two objects moving solely under the influence of their mutual gravity force fields are ellipses or hyperbolas or, under very special conditions, circles, straight lines, or parabolas. (The form and shape of the curve is independent of whether the frame of reference is located in either body or fixed in their common center of mass.)*

In this way Kepler's law is applicable not only to planets moving about the Sun, but also to satellites about a planet or satellite, or to a spacecraft simply "flying by" a planet or satellite, deflected by its gravity field. Of course a satellite will not *influence* the motion of its primary, in any measurable amount, unless its mass, relative to that of its *primary,* is a factor to be contended with. Of all the satellites in the Solar System, our Moon is the largest and most massive relative to its primary; indeed it is a factor *to be* contended with, for the Earth moves in a monthly period toward and away from the Sun by a not insignificant amount, and the Moon's path around the Sun is *always* concave toward it. The Moon's mass is 1.2% of the Earth's.

Even before Kepler succeeded in verifying that the orbits of Mars and the other planets were ellipses, with the Sun in a focus of each, he made an intuitive guess as to the nature of their nonuniform motion in their orbits. He put forth the bold assertion that *the area under the curve swept out by the line from the Sun,* not the path *along* that curve, *was uniform with time*. It would require the methods of differential calculus to verify that this statement, *Kepler's second law,* is exactly equivalent to the statement that the angular momentum of the planet about the Sun is a constant of the motion.

The general statement of *Kepler's second law* might be given as follows:

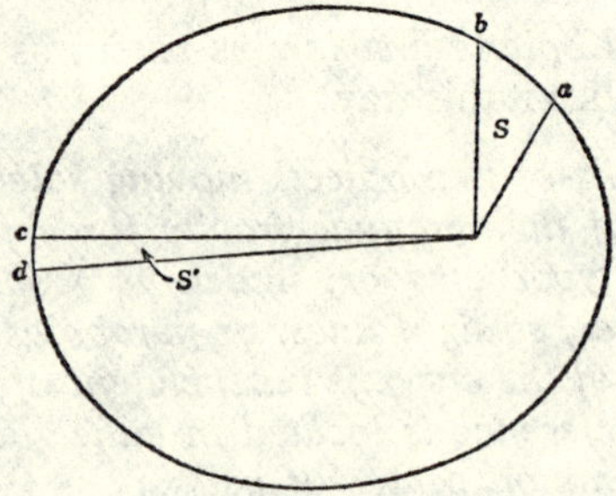

FIG. 42. Kepler's second law states that if S and S′ are of equal area then the time it takes for a planet to move from *a* to *b* must be the same as from *c* to *d*.

The time rate at which area is swept out by the radius vector which describes the path of either of two objects moving solely under the influence of their mutual gravity force as measured from an origin placed in the other object or in their common center of mass is constant. Assuming that one of the bodies is stationary this constant is equal to one-half of the angular momentum per unit mass of the moving body.

By writing the formulation this way its usefulness extends beyond planetary ellipses about the Sun to general conics about any primary.

The first and second laws go far in expressing the *detailed* motion of a planet about the Sun or a satellite or unpowered spacecraft about a planet, predicting its location, its velocity, and its direction of movement. They are the essential laws of space mechanics.

The *third law* came much later. It is concerned with the *integrated* motion—the total time it takes to complete an orbit in terms of the size of the orbit and strength of the gravity force. If the time rate at which the radius vector sweeps out the area under the curve is half the angular momentum per unit mass, then the *total area* of the ellipse divided by the *total time* to sweep out this area, called the *period,* will also equal half the angular momentum per unit mass.

Now the area of an ellipse is πab, where a and b, we recall, are the semimajor and semiminor axes, respectively, of the ellipse. We know that b can be written in terms of a and the eccentricity of the ellipse. Angular momentum is mathematically related to eccentricity in the same way but also depends upon a as well as the strength of the gravity force field, as we would expect. By direct application of algebra we get a simple relationship between the mean distance of an orbiting object from its primary, the time it takes to complete the orbit, and the strength of the force field itself. Kepler expressed it thusly: *The square of the period of a planet around the Sun is proportional to the cube of its mean distance to the Sun.* We must be careful to define terms. "Mean distance to the Sun" is neither the average computed by, say, taking a reading to the Sun every second of the year, nor is it the radius of a circle of the same area as the ellipse. Rather it is the average computed by taking a reading, say, at every kilometer along the orbital path. This will turn out to be the semimajor axis, a, in the limiting case where a reading is taken not only at every kilometer, but *at every point* in the orbit. The phrase "is proportional to" assumes that all planets belong to the same Solar System. An evaluation of the proportionality constant would yield information about the Sun's mass, or equivalently, the strength of the Sun's attractive field at some fixed distance.

Again we can make a more general statement, this time for *Kepler's third law,* for all elliptical orbits about any gravitational center, including the special degenerate cases of a circle and straight line of finite length reaching toward the gravitational center (see Figure 37):

> *The period of either of two bodies in orbit about the other is proportional to the three-halves power*

of the semimajor axis of either ellipse as measured from the centre of mass of the other. The proportionality constant is equal to 2π divided by the square root of the product of the "universal gravitational constant" and the combined orbiting masses.

In practical cases it turns out that we can consider the gravitational force as emanating from just one of the bodies. The other body finds itself in the gravity *field* of the first which is assumed to be spherically symmetric and, for all purposes, infinitely massive. The Earth can hardly be expected to influence the motion of the Sun, nor an artificial satellite to move the Earth. But for the record we ought to note that the center of mass of the Earth–Sun two-body system, about which the two revolve, is displaced 450 kilometers from the Sun's center, and Jupiter and the Sun mutually revolve about a point 740,000 kilometers from the Sun's center. This is almost 50,000 kilometers above the Sun's surface and makes one reflect on whether an astronomer at Alpha Centauri, watching the Sun exhibit its 11.9 year wobble might not begin to wonder about the Sun, whether it "too" has something going around it, and whether that something might not just possibly. . . .

Coming back to Kepler's third law, the square-rooted quantity turns out to be the *k* of Chapter Four, which we read out as the gravity field strength anywhere times the square of the distance to the gravity source. When we apply the formula to the motion of satellites in Earth's gravity, by custom we tend to express the field strength in terms of the measured acceleration of freely falling objects. For low-orbiting satellites this is 9.8 meters per second per second, which corresponds therefore to 9.8 newtons per kilogram field strength. We can avoid the proportionality constant entirely and find either the mean distance and period if one is known, and if we already know both the mean distance and the period of another orbiting satellite as we saw at the close of Chapter Two.

A fourth law

With Kepler's three laws we now have the means of describing nearly all types of situations encountered in *basic space mechanics*—the mechanics of an object moving exclusively under the influence of a dominating body, such as the Earth or Sun, assumed to give rise to an essentially spherically symmetric gravity force field, a simple form of the "two-body" or *Keplerian* problem. But suppose we are faced with the urgent question: How fast is our unpowered spacecraft actually moving at some instant when it is at some specified distance from its source planet, or the Sun, assuming again that its motion is governed to a high degree of approximation exclusively by that planet or Sun? (A trip from the Earth to Mars is considered here not as a general four-body problem, nor even a restricted four-body problem, where one of the bodies, the spacecraft, is unable to accelerate any of the three other bodies in any noticeable amount. Instead, the complexity of this problem can be reduced; we can approximate it as three two-body problems joined together, where the Earth, Sun, and Mars each play their roles, respectively, as the dominant center of gravitational attraction. The two other bodies, as well as the other planets, play out their roles in exerting *perturbative* effects. Of course, as the rocket motors are turned on and off and as atmospheric drag is encountered, and when solar radiation and magnetic forces enter, a more detailed description of the motion is eventually developed.)

When we think of speed we think of kinetic energy, which, in orbit, we continuously trade off with potential energy—except in the special case when the spacecraft is in circular motion about the gravitational center and kinetic and potential are separately maintained constant. In all cases the kinetic–potential *sum* is kept constant at all times. We always know the potential energy if we know the distance to the gravitational center. But how can we find the total energy? The easiest way to evaluate this constant total energy is to choose an especially simple location–for example, the nearest point at which a satellite approaches the Earth, its perigee–and make the calculation of potential plus kinetic energy there. At this point only, as well as at apogee, the satellite has no radial component of its motion at all–it neither moves toward or away from the center but only around it; therefore the energy calculation can be made fairly easily.

When the *Newtonian equation of motion, F = ma,* is solved for an object orbiting in an *inverse-square gravitational field,* we are able to find the r corresponding to each value of the polar angle θ. We of course work in polar coordinates because of the symmetry of the grav-

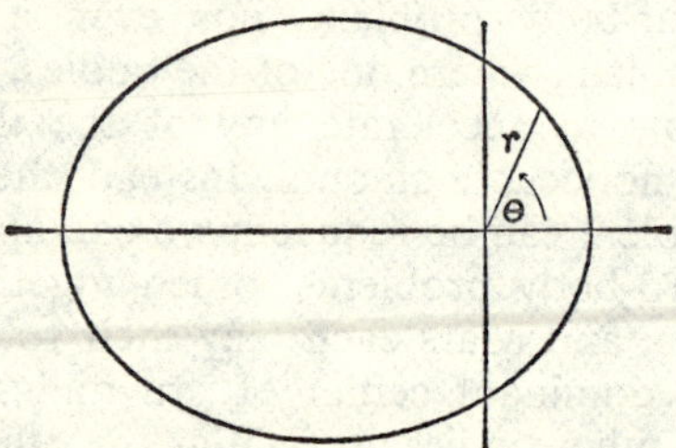

FIG. 43. Because of the symmetry of an ellipse and its properties it is nicely described by means of "polar coordinates" centered in one of the two foci. For every angle θ we can find the distance r from the focus which corresponds to it; in that way we can trace out the curve. The mathematical expression for relating r and θ is straightforward and fairly simple. For an object in orbit, the size and shape of its ellipse depends on the strength of the gravity field's pull toward the focus and on the angular momentum given to the orbiting body when it was first cast into the field.

ity force field about the origin. This polar form of the equation of motion is written in terms of the angular momentum per unit orbiting mass, or *specific angular momentum* as it is sometimes called, and k, the force field strength, as well as the shape of the orbit, described by its eccentricity ϵ. The specific angular momentum is expressible entirely in terms of k, a, and ϵ; alternatively at perigee, perihelion, or at any distance of closest approach to the gravitational center, it is merely the speed there times the distance.*

Taking these facts together we are able to find directly the total energy per unit mass of an orbiting object—the sum of its kinetic and potential energies divided by its mass. It boils down to the surprisingly simple result,

$$E = -\frac{k}{2a}.$$

This is similar to the form that we found for the total energy in a circular orbit in the note at the close of Chapter Four. There the calculation was easy because the kinetic and potential *separately* were constant. In all orbital cases, whether ellipse or circle, the kinetic energy is never great enough to overcome the negative potential energy and so the sum is negative.

The total energy of an object moving along a hyperbolic path is *positive* because the kinetic energy *exceeds* the magnitude of the potential energy even at great distances from the origin, permitting it to escape from the attracting gravity field entirely. The form of the total energy expression is the same as that for an ellipse except that now the minus sign changes to plus. Needless to say, the total energy in a parabola, the transition curve from ellipse to hyperbola, is *zero*. The mathematical equation which describes the hyperbola

* The solution to the polar form of the ellipse in terms of the gravitational–dynamical parameters is $r = p^2/k(1 + \epsilon\cos\theta)$. The specific angular momentum is $p = \sqrt{ka(1 - \epsilon^2)}$.

has the same form as that of an ellipse except that the plus sign in the parenthesis (see footnote page 111) is reversed. Also, in the expression for specific angular momentum in a hyperbola a minus sign appears within the square root, for the eccentricity will always be greater than one.

It bears repeating that, for an ellipse, a negative total energy means only that the potential dominates the kinetic, that "higher" or "lower" total energy refers merely to a less negative or more negative value, respectively, and that kinetic exactly cancels potential at every point along a parabola. Curiously enough, if the spacecraft were given slightly more speed at launch—say 12.0 kilometers/second near the Earth's surface instead of the escape, or parabolic, speed of 11.2 kilometers/second–the speed it will have "left over" far from Earth is *not* 0.8 kilometers/second but *4.3 kilometers/second.* This paradoxical conclusion comes immediately out of the law of energy conservation, our *fourth law of space mechanics:*

$$\begin{aligned}&(\text{kinetic} + \text{potential})_{\text{at launch}}\\ &= (\text{kinetic} + \text{potential})_{\text{far away}},\end{aligned}$$

or

$$\begin{aligned}&\tfrac{1}{2}(12 \text{ kilometer/second})^2 + V_{\text{at launch}}\\ &= \tfrac{1}{2}v_\infty^2 + V_{\text{at }\infty}.\end{aligned}$$

$V_{\text{at launch}}$ can be written in terms of the parabolic speed as $-\tfrac{1}{2}(11.2 \text{ kilometers/second})^2$ because the kinetic–potential sum must add up to zero in a parabolic path. Furthermore, $V_{\text{at }\infty}$ is zero. We conclude therefore that v_∞, the speed "at infinity," is the square root of the difference of the squares of 12.0 and 11.2 kilometers/second, giving us 4.3 kilometers/second.

As a general rule, if we want to be moving far away with a "residual" speed, or *excess hyperbolic speed* as it is usually called, the (relatively small) velocity increment is given to the spacecraft as close to the gravitational center as possible. Thus, as we saw above, an extra speed of 0.8 kilometers/second becomes 4.3 kilometers/second far away. We met this principle earlier in the discussion of multistage rockets. (See also Figure 28.)

But we must return to elliptic orbits and our original question: How fast is our spacecraft moving at any given distance from the gravitational center? We write the total energy expression and its kinetic and potential components,† and solve for the speed

$$v = \sqrt{k\left(\frac{2}{r} - \frac{1}{a}\right)}.$$

This is the *vis viva*, the life force, equation. It can also be used when a spacecraft is moving in a hyperbola, but in this case the minus sign in the parenthesis becomes plus.

Some space mechanics examples

1. Suppose we are in a satellite in circular orbit, moving at speed v_c about the Earth, and we fire a bullet in the direction of its motion at a muzzle speed v_c.

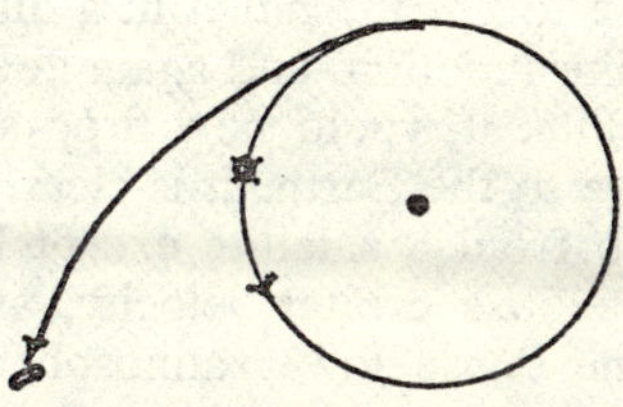

FIG. 44. A bullet fired from a satellite in circular orbit at the same speed and same direction as the satellite will move off in a hyperbola.

† $-\dfrac{k}{2a} = \frac{1}{2}v^2 - \dfrac{k}{r}.$

What do you think will be the path of the bullet? And how fast will it be moving at a great distance from the Earth?

Relative to the Earth the bullet's speed just after firing is $2v_c$. This *exceeds* the parabolic speed, which you remember is $\sqrt{2}v_c$, and so the path is a hyperbola. The perigee of the bullet's hyperbola is at the firing point, for that is the only place where the bullet's velocity is normal to the line, the radius vector, drawn from the Earth's center. The excess hyperbolic speed, or speed at infinity, is found exactly as in the example we just looked at:

$$\begin{aligned}\text{Total energy} &= K_{\text{at launch}} + V_{\text{at launch}}\\ &= K_{\text{at infinity}} + V_{\text{at infinity}}.\end{aligned}$$

It is convenient to express the *magnitude* of the potential energy at launch in terms of the parabolic speed there, as $\frac{1}{2} v_p^2$. The potential energy at infinity is, of course, zero, so now everything can be written in terms of speed. We conclude that the excess hyperbolic speed is $\sqrt{2}v_c$, which is somewhat greater than the *difference* between the launch speed and the parabolic speed needed to take something to infinity with zero speed left over.

2. If you were to fire the bullet in a direction opposite to the satellite's motion, and again at speed v_c relative to the satellite, it would now drop straight down *from rest* relative to the Earth. It is clearly not so easy to drop a bomb from a satellite except by cancelling most of the satellite's orbital velocity, and then take advantage of the dense lower atmosphere to slow it down.

Suppose we again fired a bullet from the satellite at v_c but this time pointed the gun straight down toward the Earth's center? What would its orbit look like? How close would it come to the Earth and how fast would it be moving there?

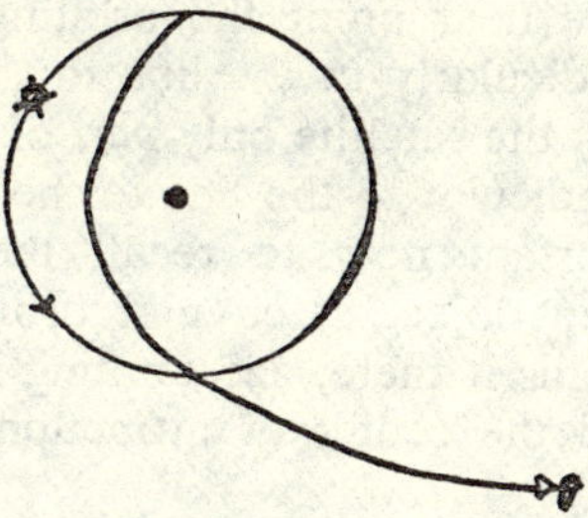

FIG. 45. A bullet fired from a satellite in circular orbit at the same speed but toward the Earth will move in a parabola, the perigee of which will be at half the radius of the satellite's orbit.

Part of the bullet's velocity with respect to the Earth derives from the gun's muzzle velocity directed toward the Earth, and part comes from the horizontal velocity the gun had in moving, as part of the satellite, around the Earth. The horizontal and vertical velocities combine to move the bullet initially 45° down from the horizontal. Thus, its actual speed, after combining both velocities, is $\sqrt{2}v_c$. This again is the parabolic speed, and so the bullet takes off for infinity with the minimal speed needed to get there, with no speed left over when it "arrives."

To find the perigee point of the parabola and its speed there, we must solve together the energy conservation equation (the fourth law) and the angular momentum conservation equation (the second law).‡ No-

‡ The energy equation at the near point, at r' from the center, is

$$0 = \tfrac{1}{2} v_p'^2 - \frac{k}{r'},$$

where v_p' is, of course, the parabolic speed at r', which we are to find.

The angular momentum equation, found at both r and r', is

$$v_c r = v_p' r'.$$

tice that at the near point the orbiting bullet moves exactly perpendicularly to the line to the Earth's center; however, at the satellite only *part* of the bullet's velocity is perpendicular to the line to the center, namely v_c. It is important now to recall the special relationship that an object in circular orbit has with the force which holds it there, accelerating it *centripetally,* that is, toward the center of attraction. This tells us that

$$v_c{}^2 = \frac{k}{r}.*$$

Putting these facts together we find that $r' = \frac{1}{2}r$ and that $v_p' = 2v_c$–the speed of the bullet at perigee is twice the speed of the satellite. This conclusion does not mean that the bullet moves in a hyperbola, since the parabolic and circular speeds at $\frac{1}{2}r$ are much greater than they are at the satellite's radius r.

3. Suppose the bullet were now fired *normally to the plane* of the satellite's orbit. What would be its path?

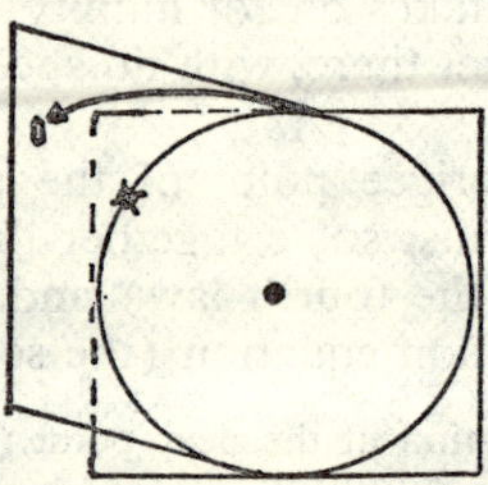

FIG. 46. A bullet fired from a satellite in circular orbit at the same speed as the satellite but normal to its orbital plane will move off in a parabola in a new plane inclined at 45° to the satellite's orbital plane.

* Found by equating the force per unit mass, k/r^2, to the acceleration $v_c{}^2/r$.

Relative to the Earth, the satellite's speed again is $\sqrt{2}v_c$ for the same reason as in Example 2–the gun's velocity through space and the muzzle velocity are perpendicular to each other. The path, again, is parabolic. This time, however, the bullet *moves out of the satellite's orbital plane* at an inclination of 45 degrees. At the point of release the bullet's direction of motion is exactly normal to the direction to Earth's center; this then is the perigee point of the parabola.

4. Let us plan to rendezvous with an asteroid that orbits the Sun in a circle at seven astronomical units from the Sun. (An astronomical unit, remember, is the semimajor axis of the Earth's orbit, or the mean distance from Earth to Sun.) What speed will our spacecraft require relative to the Sun at the start of the journey, and how long will the trip to the asteroid take?

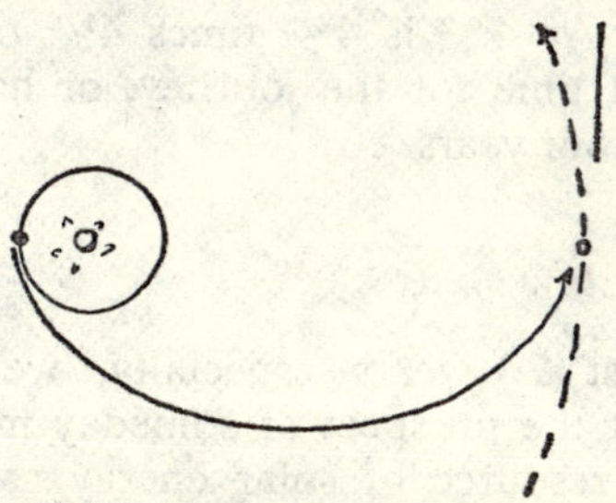

FIG. 47. To contact an asteroid moving at seven times the Earth's distance from the Sun we must produce an interplanetary ellipse which is tangent to the orbits of both the Earth and asteroid.

By aiming the launch into interplanetary space in the same direction as the Earth's movement around the Sun, we can minimize the propellant inventory. The perihelion speed relative to the Sun of the interplanetary ellipse is found directly from the *vis viva* equation.† Then we can express it, as in Example 3, in terms of a circular speed, the mean speed of the Earth around the Sun, of 29.8 kilometers/second (18.5 miles/second). As the semimajor axis of the interplanetary ellipse, *a*, is $\frac{1}{2}(1 + 7) = 4$ astronomical units, then it is easy to find how fast our spacecraft must be moving after leaving the Earth, while still in the perihelion region:

$$v_{pe} = \sqrt{7/2}\,v_c = 39.4 \text{ kilometers/second (24.5 miles/second).}$$

The duration of the journey can be expressed in Earth years by writing the period of the ellipse as

$$T = a^{3/2},$$

where the proportionality constant in Kepler's third law is simply unity, provided both units are expressed as Earth orbital units (years and astronomical units, respectively). As $4^{3/2}$ is $4^{2/2}$ times $4^{1/2}$ or $4\sqrt{4} = 8$, then the flight time for the journey, or half period of the ellipse, is four years.

Sailboats and slow boats

In the past 20 years especially, we have often thought about the prospect of someday making use of the immense resource of solar energy as a means of eliminating the chemical, radioactive, and thermal pollution which infects our Earth, and of substituting a renewable resource for fuels which are rapidly being depleted as the demand for energy continues to increase. Solar energy is a very realistic source of power for driving space vehicles in interplanetary missions.

† This is $v_{pe}'^2 = k\left(\frac{2}{r} - \frac{1}{a}\right)$. In this case the *k* refers to the Sun's gravitational force field strength at a specified distance from the Sun, for example, at one astronomical unit.

Suppose we were to suspend a black sheet, a square one meter on a side, in space facing the Sun and just outside the Earth's atmosphere. How would we find the radiation force on this sheet? We do know that the *solar radiation flux* has been measured as 1400 joules per second, that is, 1400 watts, per square meter. This is the same as 20,000 calories per minute, just enough energy in one minute to heat up 200 grams of water from the freezing point to the boiling point. In ten hours 120 kilograms of water could be heated through that temperature range.‡ This is why there is so much talk about the Sun solving most, if not all, of our energy problems—eventually. If that sheet were part of an immense orbiting radiation, absorption, and conversion plant one kilometer square, it could, even at 70% efficiency, deliver about 1000 megawatts, the equivalent of a rather large nuclear or coal-burning power plant. This amount of power could fill the needs of an industrial city of several hundred thousand inhabitants.

But let us return to our one square meter black sheet hanging rigidly alone in space. This radiant energy cannot become absorbed in the black sheet without producing a continuous force on it, and this force can be utilized directly for interplanetary propulsion.

It is really very easy to calculate force, or pressure, from radiant energy, for, by Einstein's relation, radiant energy is mc^2 and momentum is mc (c is the speed of light and m is the mass of the *photons* comprising a given amount of radiation). It is awkward to think of light as a rain of particles, photons, when we are so used to thinking of waves when we speak of light. Yet light behaves exactly as we would expect particles to behave, even while they continue to be waves. His *photoelectric theory* explaining the particle nature of light won Einstein the Nobel Prize, quite apart from his work on the theory of relativity. So we can imagine the

‡ We assume here that the absorbed heat is not lost through radiation by the water.

light—or totality of *electromagnetic radiation* of which light is but a small part—striking the black sheet like a rain of bullets which become absorbed and transmit a force. How is the force transmitted? The momentum changes from the value it had on its way from the Sun, to zero at the sheet. And this happens in a certain time. The shorter the time, the greater the force. The greater the momentum change at the sheet, the greater the force.

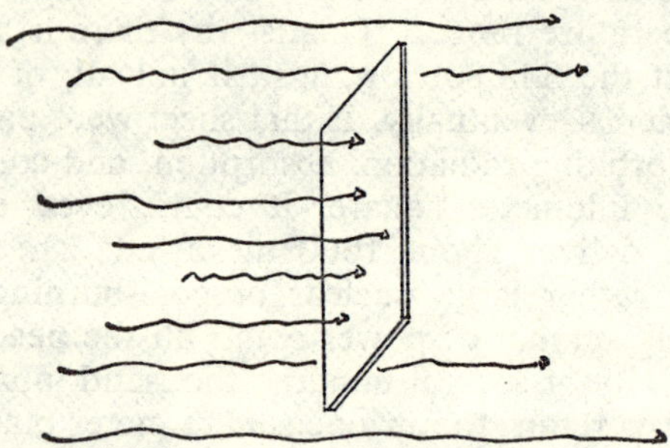

FIG. 48. The loss in momentum per second of the electromagnetic radiation which is stopped by the absorbing sheet is the total force imparted to the sheet.

As we just saw earlier, dividing the energy by the speed of light gives us the momentum of the beam, which the light loses when it hits the sheet. If we divide the power in a one square meter beam (that is, energy per second, 1400 watts) by the speed of light (300 meters per microsecond) we get a change in momentum per second, or *force,* of five micronewtons [300 meters per microsecond] (or one-half dyne).

This is not a terribly large force, but it is quite respectable compared with the solar gravity force on a thin piece of plastic of that size. To make the radiant force twice as respectable change from a black sheet to a silvery white one. Take some very thin mylar and coat it with a brilliantly shiny aluminum surface, only a few wavelengths of light thick. The force will now be *ten* micronewtons or one dyne. Of course we cannot make use of the energy of the sunlight anymore, to run electrical equipment for example, for we sent the beam on its way, back to the Sun.

A square meter of thin, aluminum-coated mylar might be expected to have a mass of about one gram, so that an acceleration of one centimeter per second^2 could be anticipated. Although this is only a tenth of a percent of Earth's surface gravity, it is *1.7 times* greater than the Sun's gravity at the Earth's orbit. Thus this sheet of aluminum and mylar is a prime candidate for a *solar sail.*

A good "figure of merit" of a solar sail is its *lightness number*–the ratio of radiation acceleration to solar gravity. The intensity of radiation and the gravity force field each obey the inverse-square law–they grow weaker together as we move out from the Sun, or stronger together as we move in.

The scientists and engineers at the Jet Propulsion Laboratory of the California Institute of Technology, the organization which has achieved one success after another in sending space orbiters and landers to the Moon and planets, have solar sails now on their drawing boards. They are currently mapping out missions across the Solar System with small payloads. One of their missions to Mercury, complete with a sail, structure, and payload, has a lightness number of 0.16.

It is a misconception to presume that a lightness number greater than unity is a must in "pushing out" beyond the Earth's orbit. On the contrary, we are all moving about the Sun already in an almost circular orbit at one astronomical unit. Once the sail has been placed in orbit (by other means than solar radiation propulsion of course), a tilt will produce a slight propulsive force *in the direction* of orbital motion, and the sail will spiral *out* into ever-widening arcs. Tilt the sail

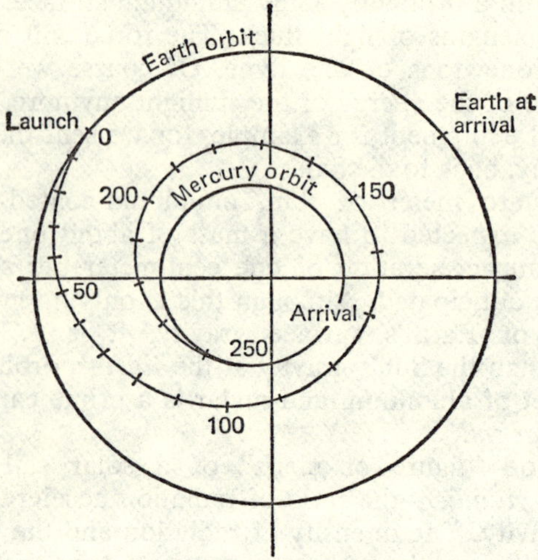

FIG. 49. A trip to Mercury by means of a solar sail starting from an orbit about the Earth, as planned by the Jet Propulsion Laboratory. The time of flight is 250 days.

the other way and radiation force acts as a *drag* force, spiraling it *in* toward the Venusian and Mercurian orbits. As the propulsive thrust is not a microscopic fraction of the solar gravity but comparable with it, the powered trajectory leading to the final destination will not resemble the groove of a phonograph record, but rather a watch spring where the destination is reached in only two or three circuits.

To Mars and beyond

In addition to the solar sail, an ion-powered rocket, with *thrust-to-mass ratio* programmed to decrease with solar distance in the same way as solar gravity, would move in a spiral of constant pitch. This so-called *logarithmic spiral,* aside from being a simple path to

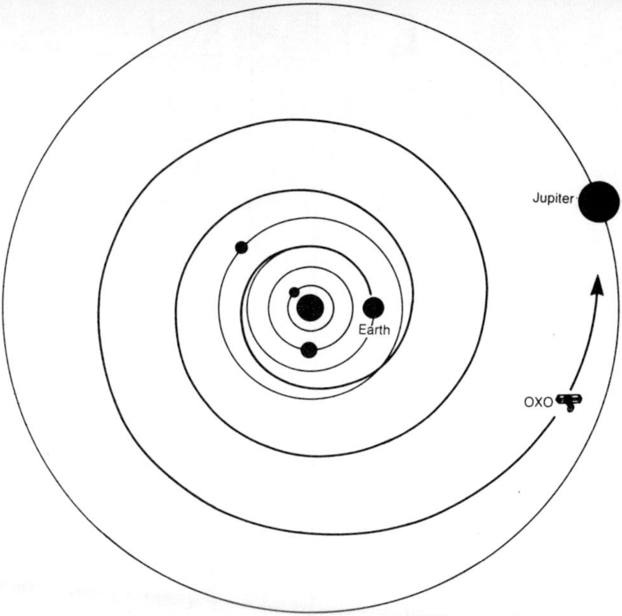

Two ways to reach the orbit of Jupiter. In the top view the hypothetical spacecraft, called OXO for Orbiting X-ray Observatory, is launched in an ellipse, cotangent to the Earth and Jovian orbits. Following Kepler's third law, where the square of the period is proportional to the cube of the ellipse's major axis, the trip takes two and a half years. A small retrothrust can bring the spacecraft close to the Sun. In the lower view the OXO is in a powered trajectory where either an ion engine or reflected sunlight can provide the thrust. Spiral journeys can be made in relatively short time.

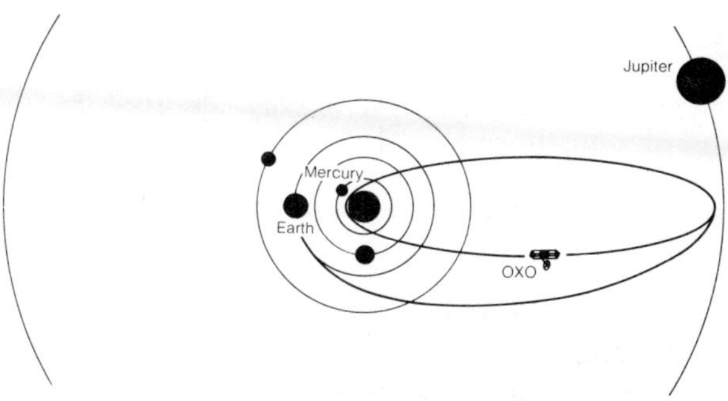

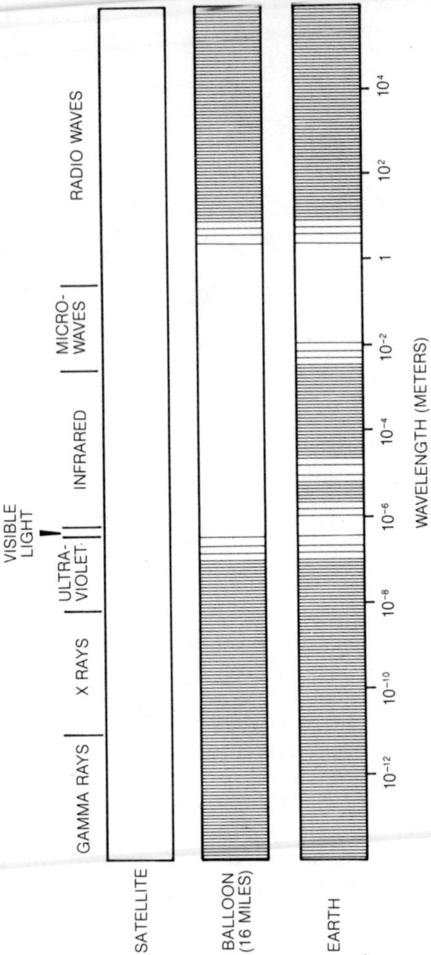

Radiation from space is heavily absorbed by the Earth's atmosphere. At balloon altitudes the infrared region becomes accessible but at the altitude of space observatories the entire electromagnetic spectrum is visible.

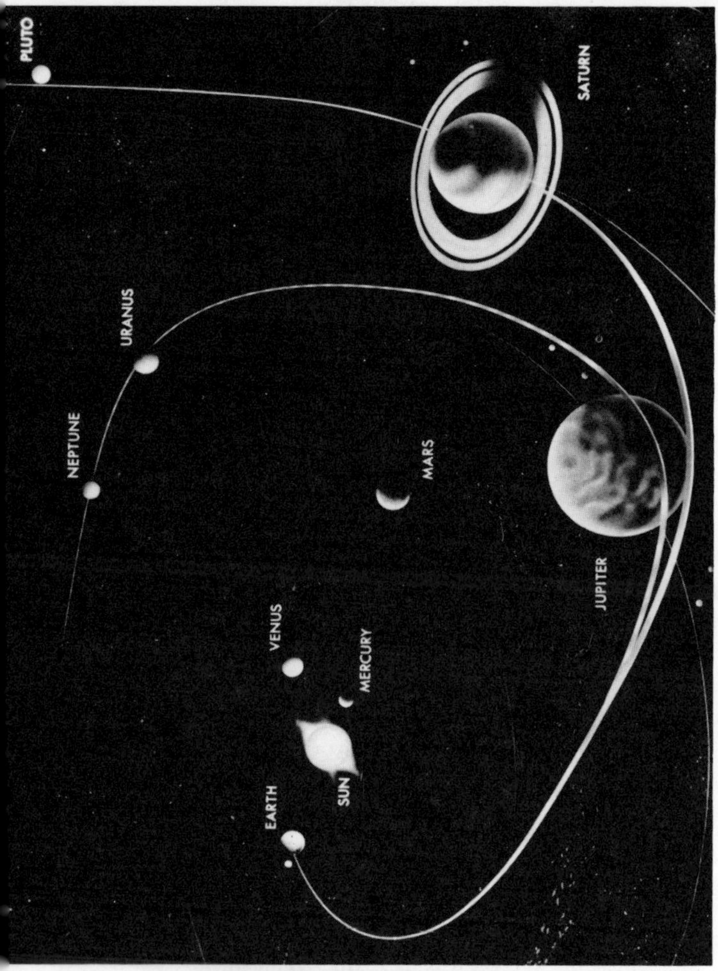

The gravitational field of a planet, especially Jupiter, can be used to both change the direction of flight *and* add propulsion energy to an approaching spacecraft. In this case trips to the outer planets can be contemplated when they are lined up at the right places. Mariner 10 was able to make use of the principle by traveling to Mercury after being deflected by Venus. (*IPS photo*)

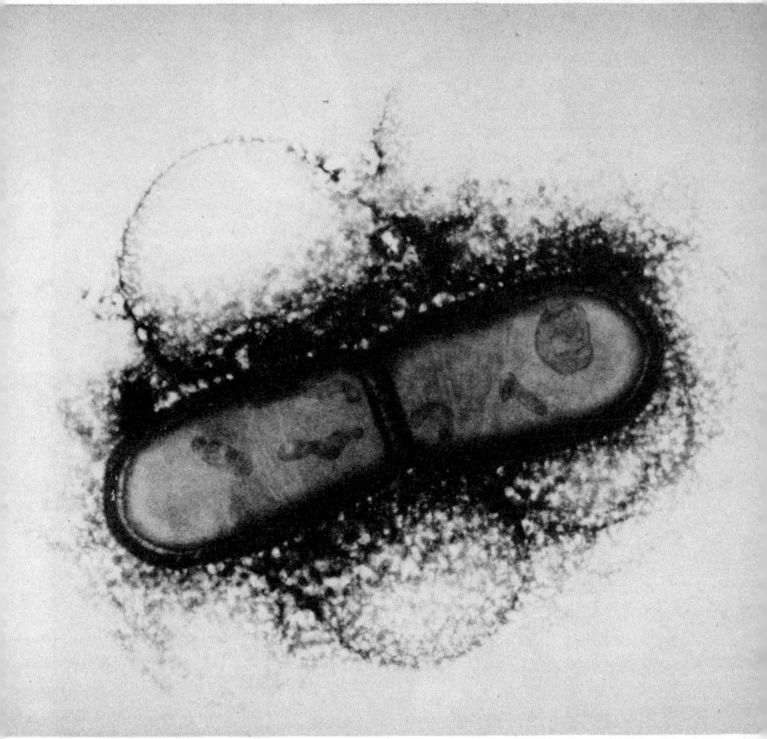

Two biologists at the Ames Research Center of the National Aeronautics and Space Administration found this fissioning organism thriving in an environment as alkaline as Jupiter is expected to have—ten times more alkaline than the previous maximum level thought possible for growth of living organisms.

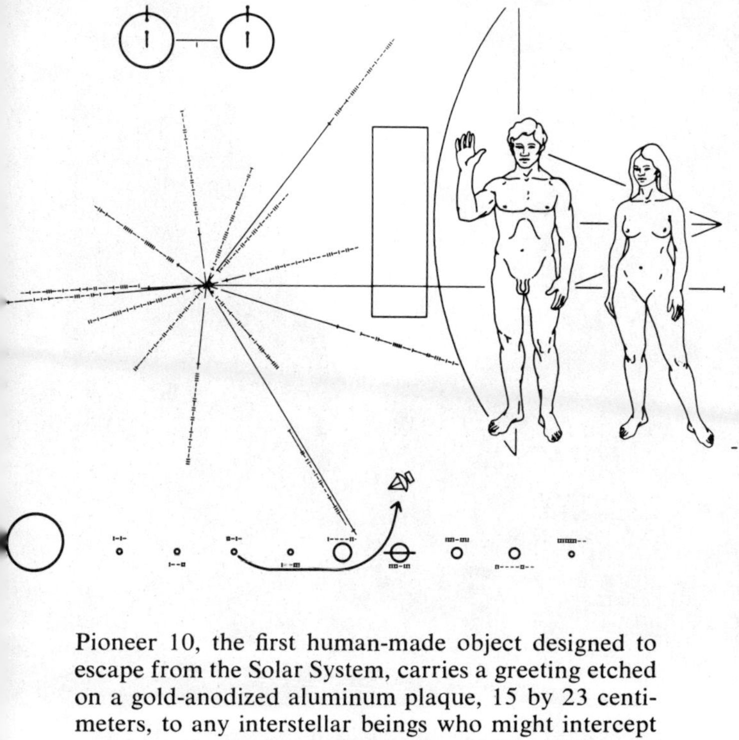

Pioneer 10, the first human-made object designed to escape from the Solar System, carries a greeting etched on a gold-anodized aluminum plaque, 15 by 23 centimeters, to any interstellar beings who might intercept and have the intelligence to understand it millions of years in the future. (*IPS photo*)

Design for a solar sail. The sail is constructed of very thin mylar coated with an aluminum reflective layer. It is about one kilometer across and requires an elaborate system of guy wires to keep it rigid. The sail can be tilted to produce a strong solar radiation force component either in the direction of orbital motion or opposed to the orbital motion, thereby spiraling away from the Sun or toward the Sun, respectively.

The problem of deploying a solar sail from the nose cone of a rocket is a difficult one, as it cannot be easily unfolded. This concept—the heliogyro—utilizes very long strips of mylar unrolled from drums and maintained rigid by rotation of the vehicle. The angle of tilt can be adjusted by tilting the axis of each drum.

Gerard O'Neill's concept of two orbiting space colonies. Solar radiation would be the power source, and lunar raw materials would be used for construction. Each cylinder is 32 kilometers long and 6.4 kilometers in diameter, providing a home, work, and recreational climate for between 200,000 and several million persons. Earth-like gravity would be produced by rotating it every 114 seconds. The large movable rectangular mirrors on the sides direct sunlight into the interior to provide seasonal and day–night changes. The teacup-shaped objects in a ring at the ends of the cylinders are agricultural stations.

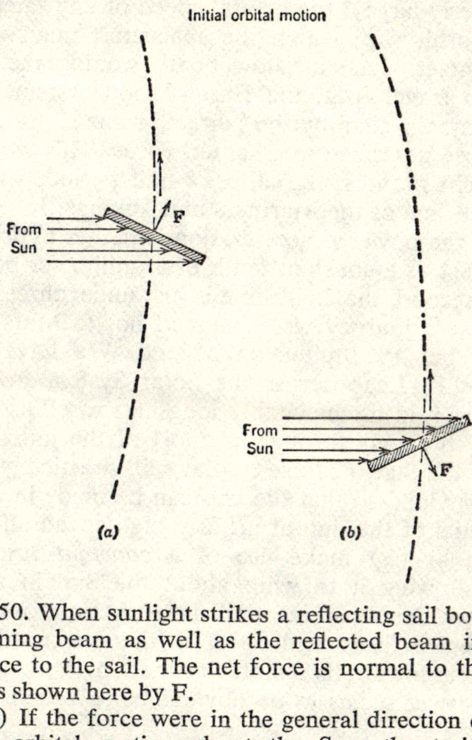

FIG. 50. When sunlight strikes a reflecting sail both the incoming beam as well as the reflected beam impart a force to the sail. The net force is normal to the sail and is shown here by F.

(a) If the force were in the general direction of the sail's orbital motion about the Sun, the trajectory would be a spiral moving the sail further out from the Sun.

(b) If the force were in a direction *against* the orbital motion, it would be, in effect, a drag force which causes the sail to spiral in toward the Sun. One can construct sails out of mylar supporting an aluminum coating of a thickness measured in wavelengths of light. The mylar is five times thinner than household plastic wrapping.

work with on the computer, has the advantage of matching reasonably well (so long as the spiral's pitch is not overly large), the orbital speed of any satellite in circular orbit with which the spacecraft may wish to make contact. Such a "slow boat" would take much longer to travel from the Earth–Moon system to the Martian system than by the "direct" route of the cotangent ellipse described in Chapter Three. This would be about eight months, the ellipse's half period, which is the travel time of the Mariners and Vikings. In a spiral in which the powered acceleration, whether by ionic or solar thrust, is a constant tenth of a millimeter per second per second, the flight time is just under three years.

Suppose a journey were planned not to Mars or its satellites but to Jupiter or beyond. We have three choices: (1) Leap across the Solar System from the Earth's orbit in the inevitably long narrow ellipse moving ever so slowly throughout most of the journey far out from the Sun; (2) use a solar sail, drastically limiting our payload so that the trip can be made in one or two circuits of the Sun at a fairly high speed all along the way; or (3) make use of a *constant-thrust* ion rocket, allowing it to whip about the Sun in a continuously widening spiral until the rocket motors speed the spaceship faster and faster in the weakened solar gravity field toward its remote destination.

The answer seems to be obvious as to which choice to make:

> The constant-thrust rocket will get us there soonest. As the solar field fades away the acceleration of the rocket will continue to build up as less and less propellant remains in the tank. And even though the acceleration is tiny, over many years the rocket will reach fantastic speeds.

This may well be the way to the stars. Only one small question remains: How will we ever manage to slow it down when we get there?

CHAPTER SIX

How to Stand Up in Space Without Even Trying . . . by Taking Along Your Own Gravity Field

A funny way to design a building

It's so hard to find a really fast-moving and fast-accelerating elevator these days. Like so much in life, perhaps too much, their slightest departures from perfection are electronically controlled away. Recently I visited the very impressive ultramodern buildings of the University of Odense–Hans Christian Andersen would never have recognized the city of his birth–and walked into a very large elevator suitable for carrying laboratory equipment, robots, and people. Like most elevators in Europe the car had no gate so you could see and touch the large gray stone building blocks of the wide shaft just in front of you. Suddenly, after pressing the button marked "3", the whole building began to move quietly down; the elevator and I were so still you could see a quantum jump. After a few seconds a door marked "3" came to rest in front of me. I walked through, and thought, "What a funny way to design a building!" There is a large Japanese elevator company which advertises its electronic skills in international magazines by showing a French franc balanced on its edge on the elevator floor at ground level and demonstrating that the coin stays balanced even after the elevator meets the top floor of a large office building. The same principle as in Odense.

But did you ever stand on a really fast elevator, one

that must take you fifty or sixty floors up in a hurry? Suddenly it slams on its brakes. Your stomach keeps right on going. In that brief moment your feet are somehow still on the floor but not pressed down nearly as much as before. You have lost some weight, that much is sure. If you had brought a bathroom scale and stood on it, you could confirm it at once.

The elevator then begins moving down. Faster and faster and then—brakes on again. This time your stomach presses down against you. It wants to continue moving but your feet say no. Well, in fact, your feet want to keep moving, too, but the elevator says no. (Or the elevator wants to keep moving but the cable . . . Oh well.) If you again stood on the bathroom scale for that fraction of a second, you would see for yourself that now you really *weigh more.*

Both Galileo and Newton were aware of the resemblance between accelerating reference frames and gravitational fields. Drop two balls, a wooden ball and a steel ball, and they both fall to the ground together. Galileo said that the acceleration—the rate at which the velocity changes with time—is exactly the same for each. But why? Newton explained it this way: The balls are pulled down by an attractive force of gravity that is proportional to mass, and then . . .

Mass? What is that?

N: *The quantity of matter in a body.*

You mean the steel ball is more massive, is harder to lift . . .

N: *That's the point. The pull of gravity is greater for the steel one. Therefore it is more massive.*

But if the Earth pulls it down harder, explain why it does not fall down faster.

N: *Because its large inertia—or inertial mass—resists the pull by just the right amount so that it moves down exactly at the acceleration of the wooden ball.*

You mean the gravitational mass, which determines the force, exactly equals the inertial mass, which determines the acceleration?

N: *Precisely.*

But suppose I were out in space, far from any sources of gravity, and accelerating in a rocket at 9.8 meters/second^2. First of all, where is the floor?

N: *The inside cabin surface furthest from the direction of the rocket's acceleration.*

Do you really mean "direction of its motion"?

N: *Certainly not. The* acceleration *might actually be lowering your velocity if you happened to be moving "backward." In any case, you might be moving at one velocity relative to the Solar System, and quite another relative to the center of mass of the Galaxy. Who knows where the absolute reference frame really is?*

If only you had remembered to say *that* in your *Principia!*

N: *This "floor" is where we can stand up in space . . .*

Without even trying! At last we know which way is "up."

So, if I release the steel and wooden balls . . .

N: *They will "accelerate" to the floor together.*

But they are really moving at constant velocity, the velocity they had inside the rocket at the moment of release.

N: *And the floor will accelerate up to meet them together.*

Then the accelerating rocket in "gravity-free" space behaves exactly as if it were sitting on the launching pad on the Earth.

N: *Almost exactly.*

Why just almost?

N: *For one thing, the gravity force field of the Earth points straight toward the Earth. If you hang plumb-lines at various places inside a room, they are*

FIG. 51. The spaceship accelerates in gravity-free space at a constant 9.8 meters per second per second in the direction of its nose. The astronaut (a physics professor) stands up in space as if he were in his lecture hall on the Earth. When he releases two balls they no longer are accelerating, and so the spaceship's floor overtakes them. He writes in his notebook: "They fell to the floor together, at a constant acceleration of 9.8 meters per second per second," a conclusion that convinces him he is really on Earth. (From "Gravity," G. Gamow, © 1961, *Scientific American*.)

not quite parallel. Furthermore, the force of gravity near the floor is a tiny bit stronger than near the ceiling. Inside the accelerating rocket, on the other hand, the direction of the "force field" is the same everywhere. And the force "strength" is exactly the same at the ceiling as at the floor. So there you have the distinction between an accelerating rocket in space, and one sitting on its launching pad on Earth. Another thing that bothers me in the accelerating rocket is the apparent acceleration, backwards, of the rest of the Universe.

We had better first pull down the window shades and then stick to making measurements in a space that is far, far smaller than the dimensions of the Earth.

N: *Like an ordinary room, or rocket interior. Let us be practical. In that case, in every respect, the rocket on the launchpad on Earth and the accelerating rocket in space are equivalent. This, in fact, is the famous* principle of equivalence.

So if we agree that they are equivalent, then it must follow that the weight measured in an accelerating frame—an accelerating rocket, an accelerating elevator, and so on—is the *true* weight and not just some kind of apparent weight, or effective weight.

N: *In every case it is the force of an object upon its support.*

But suppose an object, say an empty rocket casing, is *freely falling,* that is, accelerating toward the Earth's center in its gravity field. Is it weightless?

N: *If it falls* in vacuo, *as you implied, it is. We could have looked at it this way: In falling—and remember, moving in a circular satellite orbit is also falling—the object's own reference frame accelerates toward the Earth at 9.8 meters/second^2. If it were not for the Earth, this affect alone would have produced the condition for an equivalent gravity field directed opposite to the acceleration. But the Earth* is *there, with an opposing gravity field, and the two just cancel. Inside the freely falling rocket you would float about, just as you would if the space capsule were moving at constant velocity in interstellar space.*

This would mean that if you *powered* a rocket to accelerate, just above the Earth, straight down at two times 9.8 meters/second^2, or 19.6 meters/second^2 . . .

N: *You would stand* up, *that is to say,* "down," *with your feet toward the stars and your head toward the ground. And you would weigh, in magnitude, just what you did when you stood on the ground.*

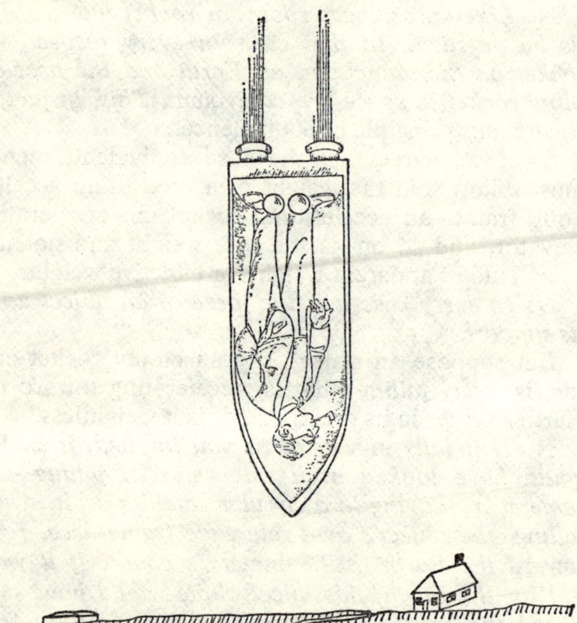

FIG. 52. Someone who knows which end is up.

Your rocket thrusters would have to produce the equivalent of only one unit of 9.8 meters/second^2, of course.

N: *Of course.*

Does this mean that if a fellow jumps out of a plane so that he accelerates down at 90% of 9.8 meters/second^2, then he weighs just 10% of his normal weight when standing on the ground?

N: *Yes, the air partly supports him.*

Then if he opens a parachute and moves down at a high constant velocity, his weight must be his normal weight, even while he is moving.

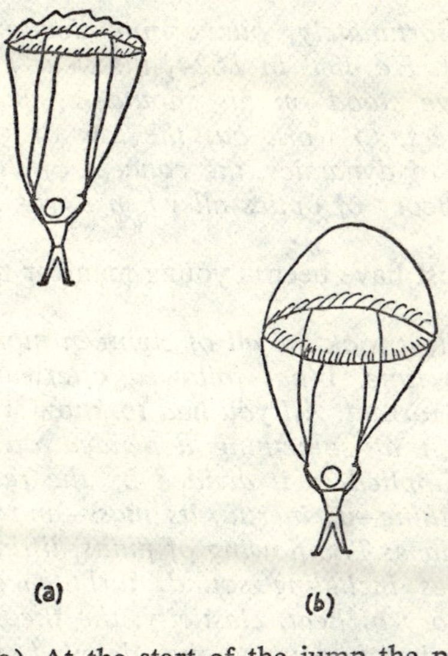

FIG. 53. (a) At the start of the jump the parachutist accelerates down at 90% of the gravity force field strength *g*. His weight, opposing the supporting force of the air, is 10% of his normal weight.

(b) Eventually he moves down at constant velocity; as he no longer is accelerating his weight is the same as it would be at rest on the Earth, or in the plane—his normal weight.

N: *Yes, because now the air gives him full support–holding back his acceleration completely.*

That must be the reason the ancient philosophers had so much trouble understanding Galileo; well, they didn't really want to understand him.

N: *Galileo tried ever so hard to explain to them that you just cannot hide the truth about the nature of falling bodies under the little smidget of air resistance which affects the steel ball so slightly and the wood ball only a bit more. When he dropped objects from the Pisa Tower–as many people say he has done–they simply refused to look.*

Have you discussed all of these things with Galileo?

N: *Unfortunately, our communication was not very good. He died in 1624, the year I was born. But I have stood on his shoulders, and those of many others, to work out the law of gravity, the principles of dynamics, the concept of the calculus, and the theory of optics all when I was still a very young man.*

You must have been a young man for many many years.

N: *Well, it took me all of eighteen months; I was 26 at the time. What followed afterward was an onrushing torrent. All you had to know was that the acceleration of something depended on the force that was applied to it divided by the resistance of that something–its inertia, its mass–and everything fell into place: The flowing of fluids, lifting by pressure, waves including sound, turbulence, the dynamic theory of heat, elasticity, the theory of electricity, and, of course, the detailed motion of planets, satellites, and comets. Would you believe that one of my hardest problems was to prove that the billions upon billions of tiny masses which compose a spherically symmetric mass–that is, the combined effects of all these tiny masses–affect other bodies as if all the masses were squeezed down to a point at the exact center. Well, people tend to take such things for granted; now that the work is done they say "it is obvious." But it was not at all obvious to me. You would think–that is, I would think–that as the masses up front would have a much greater gravitational effect than those in the rear, because of the inverse-square law, then the center of gravity of a sphere–its effective point of attraction to other bodies–might be pushed ahead of the geometric center. But I proved that this just was not so. And at least I did not have any of those rascals who plagued Galileo ready to bring the wrath of God down on* my *head.*

There is one thing I cannot quite understand, however.

What's that?

N: *Just what does it mean to "jump out of a plane"?*

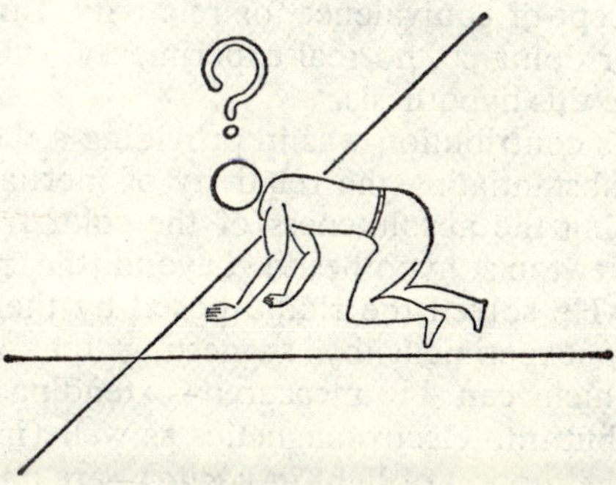

FIG. 54. Jumping out of a plane.

The principle of equivalence according to Einstein

This general conclusion that observations carried out in an accelerated cabin and in a cabin at rest (or moving uniformly) in a "real" gravitational force field are equal is what we call the principle of *equivalence.* To say it another way, a body's gravitational mass is the same as its inertial mass. But our discussion has so far been limited to mechanical phenomena: The laws of mechanics–swinging pendulums, colliding particles, flowing water, and so on–are identical in each frame of reference, the gravitational frame and the accelerating frame.

We return now to an earlier discussion: Comparing mechanical phenomena in different *constant-velocity* frames–"Galilean" or "inertial" frames–we find that they too are equivalent in that the laws of mechanics are identical in each. No experiment in mechanics could be performed to prove otherwise. No preferred reference frame could be found that stands out above the others, for all would be indeed equivalent. This was known in Newton's time and even in Galileo's time. Newton tried to simplify his world model by calling the reference frame of the Solar System, or Sun, a preferred frame, the absolute frame, thus upsetting the whole concept of equivalence, or relativity. But he had no basis for doing so, no real experimental evidence to substantiate this hypothesis.

Einstein's contribution was in providing a theoretical basis for substantiating the relativity of inertial frames (thus denying the absoluteness of the solar frame implicit in Newton's hypothesis) beyond the realm of mechanics. He solved the riddle posed by the Michelson–Morley experiment by suggesting that *all* phenomena which can be measured–extending beyond mechanics but into electromagnetics as well (including, of course, the motion of light *in vacuo*)–are unchanged when shifting from one inertial reference frame to another. Stated in the formal and precise language of science: All physical laws are invariant under a Galilean transformation. This, as I have discussed earlier, is the essence of Einstein's restricted or special relativity theory (restricted in the sense that the moving reference frames have zero acceleration).

In examining the equivalence, the mechanical equivalence, between accelerating reference frames and non-accelerating frames placed in gravitational fields, the next direction to proceed in seems to be obvious, at least it was so for Einstein: If the laws of mechanics are equivalent, what then about *all* the laws of physics? What about Maxwell's laws of electromagnetics? If light moves in a special way in an accelerating frame, how then would it move in an equivalent gravitational force field?

We do not really have to shoot laser beams about in an accelerating rocket to convince ourselves of the expected outcome. We can make a "thought experiment."

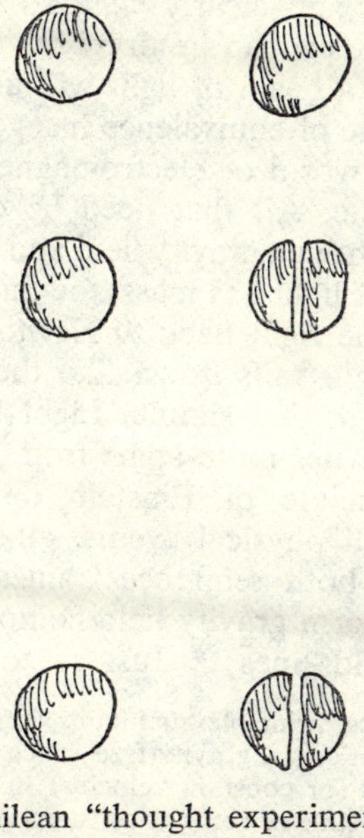

FIG. 55. A Galilean "thought experiment." If the falling body breaks into two smaller parts, the parts move down as one. There is no basis for expecting that each falls half as fast, which was Aristotle's prediction.

This is the kind of approach that Galileo used in order to "prove" that heavy and light objects fall to Earth together: Imagine a falling object breaking suddenly into two or three pieces. Should they now instantly move at much lower speeds? If so, why? Increased air resistance? Then do the whole (thought) experiment in a vacuum!

Einstein's thought experiment went as follows: Imagine a laser light beam crossing the cabin of an accelerating rocket. (Einstein's work predated lasers by fifty years, but never mind.) The rocket, as usual, is accelerating in gravity-free space. To an outside observer the beam moves, of course, at constant velocity. The floor of the rocket accelerates up to it and so, within the rocket, the beam clearly *curves toward the floor.* What other inference can he draw than: In an equivalent gravity field a beam of light will also curve down. Thus the principle of equivalence has gone beyond mechanics into the world of electromagnetics making the principle complete. All that need be demonstrated is the curving of light in a gravity field and then. . . .

But hold on. If light has mass, the quanta have mass, then we can come right back to Newtonian mechanics and say: The light falls down like the steel or wood balls. But it is not that simple. Light has electromagnetic wave properties quite apart from mechanics. The equivalence principle of Einstein demands that all measurements, all physical events, all properties must be the same in both reference frames (assuming we could find a uniform gravity field composed of parallel and constant field lines).* Just as the invariance of

* The equivalence principle maintains, of course, not only that accelerated motion in gravity-free space is exactly equivalent to being at rest (or constant velocity) in a uniform gravity field, but also that a state of rest (or constant velocity) in gravity-free space is indistinguishable from a state of free fall in a uniform gravity field. Tests of falling bodies over the centuries have confirmed that all bodies, no matter what their size or internal

FIG. 56. In the rocket the light beam appears curved as the floor accelerates up to it. Outside the rocket the beam's path is, of course, a straight line. By Einstein's principle of equivalence, light in a gravity force field of strength matching the rocket's acceleration ought also to be deflected by a like amount. (From "Gravity," G. Gamow, © 1961, *Scientific American.*)

composition, fall with precisely the same acceleration in a given gravitational field. It was only recently that the equivalence principle was tested for separate bodies held together only by their self-gravity—the Earth and Moon in the field of the Sun. If a difference in gravitational acceleration were present, there would be a slight shift of the Moon's orbit about the Earth. By laser ranging—reflection of a laser signal from Earth by fused-silica corner reflectors placed on the Moon—the shift, if it did occur, must be below the probable error of 30 centimeters. This was small enough to consider the equivalence principle as having been fully confirmed, at least for its gravitational–mechanical aspects.

physical laws in transferring between inertial reference frames required that space and time be transformed, then it also follows that in transferring between accelerating/gravitational frames space and time are transformed here as well. We can say not merely that light is attracted by a gravity source, but also that the source distorts the space around it, a space whose "coordinates" are laid out by the path which a beam of light will follow. Considering light as a collection of mass quanta falling in a gravity field will simply not yield a theoretical value of the bending equal to the measured value.

But how can you measure this bending effect? You certainly cannot send a beam of light across a room and then turn Earth's gravity on and off to observe the (microscopic) deflection. It is uncanny how nature itself provides the means for us not only to observe the deflection of light with and without a gravitational field, but to provide us with a field many hundreds of times more powerful than the Earth's.

With the development of the great astronomical telescopes the positions of the stars in the heavens have been found with great precision, especially if, for a group of stars, the relative positions from one another are to be established. One tenth of an arc second is certainly reasonable. If the air is calm and many observations are made, the accuracy can approach one hundredth of an arc second. Suppose the Sun should move in the region of the Earth's sky that is occupied by a group of stars whose relative positions were known rather accurately. Then, according to the general theory of relativity–resting on the foundation of the principle of equivalence between accelerating frames and gravity fields–stars just outside the Sun's rim would appear displaced away from the Sun as the Sun "attracted" the starlight. The displacement should be 1.75 arc seconds. There is a slight problem, however. You cannot see any of the stars in the sky during day-

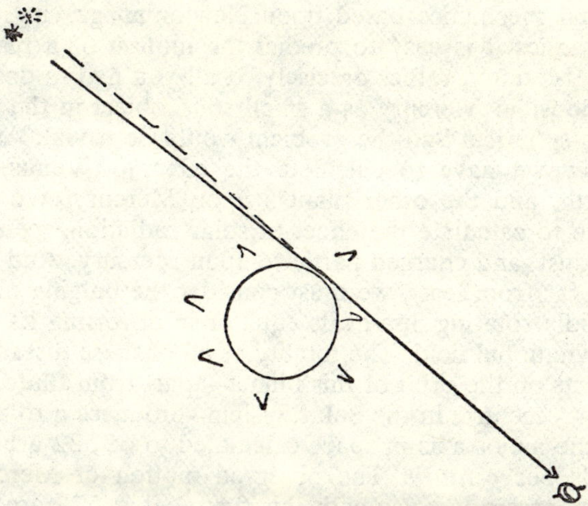

FIG. 57. The space surrounding strong gravity fields is markedly altered by the field. Thus, a light beam passing near the Sun's rim would appear in a slightly different place in the sky, as viewed from the Earth, from where it appears when the Sun is not there.

light, least of all close to the Sun, except on one occasion—when the Sun is in total eclipse. It was on such an occasion—a total eclipse of the Sun in 1919—that the British sent an astronomical expedition to the Principe Islands of West Africa. The measured displacement of stars near the solar disc was 1.61 arc seconds with an estimated error of 0.30 arc seconds. This was close enough to give a reasonably good verification of the equivalence principle for electromagnetic waves.

There were other observations which provided confirmation of the general theory based not so much on the displacement of a light beam as on other aspects of the distortion of space–time in the region of intense gravity field strengths associated with such a displacement. For example, in the theory of celestial mechanics based upon Newtonian gravitational dynamics it is easy to predict the motion of a planet, say, Mercury, rather precisely. If all you had to do was to consider Mercury as a small solid object in the field of a spherical Sun the problem would be trivial. As we know we have to calculate the effect of Venus, the Earth, and the other planets upon Mercury. We also have to calculate the effect of solar radiation, meteoritic dust, and charged particles upon Mercury. And last, but far from least, we must consider the bulging of the rapidly rotating Sun at its equator in distorting its own gravitational field. The totality of all of these disturbing effects on the orbit of this planet—apart from Pluto, the most eccentric in the Solar System—produces a rotation of the major axis in space calculated to be 532 arc seconds per *century*. The observed motion of Mercury, alas, does not agree with this figure. It is 575 arc seconds. If one were not a scientist one might be willing to take this loss of 43 arc seconds—or, more exactly, 42.56 ± 0.94 seconds—rather lightly and move on to more important matters: a comfortable easy chair, an evening with Truffaut, turning UFO's into IFO's, or Guinness, taking us through hyperspace, or Brando, contemplating the wisdom of seven galaxies amid the ruins of his planet, Krypton.

Living with an unbalanced account is a way of life, on an individual basis all the way up to the highest echelons of government. But for an astronomer such an error can wreck the very foundations of the Universe. And so it did. Observatories spent a portion of the nineteenth century's final years in searching for Vulcan, the planet that "obviously" had to exist within Mercury's orbit to provide the necessary perturbative force. But if Vulcan existed it must have been invisible, for it was never found. In the twentieth century Einstein's theory entered the picture: In gravitational fields space–time is distorted. The stronger the field the greater the distortion. And the effect on Mercury's orbit? An advance of its perihelion—of exactly *43.03* seconds per century!

This result, together with the solar eclipse data, signaled a possible breakthrough in dealing with one of the great cosmic mysteries: Where does space end? With neither a beginning nor an end, space slowly, ever so slowly, bends back upon itself, enclosing, as it does, the myriad upon myriad of galaxies. It is not their gravity which distorts space, after all; *it is our perception of the world as a neat three-dimensional package in straight lines and squared-off corners that is the distortion.*

Each year more and more scientific data appear to give greater and greater plausibility to Einstein's theories. Even the major axis of the Earth's orbit rotates slightly apart from other effects—4.6 ± 2.7 arc seconds per century, observed, and 3.8 arc seconds, calculated. In the age of interplanetary space vehicles, microwave radiation was transmitted across the Solar System and made to pass through the intense gravitational force field very close to the Sun. Einstein's prediction was again confirmed. The shifting of emitted spectral lines in the direction of less energetic quanta from stellar sources in intense gravitational fields is another manifestation of the theory. And in the extreme case of fields so intense that the quanta cannot even get out, we discover a Universe of an entirely different sort, a Universe peppered with *black holes,* beyond the reach of telescopes and probably also of space ships with any hope of returning.

The Space Wheel

The next time you are on a roller coaster, or riding in a car on a hilly road, check your stomach as you go soaring around the top of a high curve. Your stomach, along with the rest of you, wants to keep shooting off in a straight line. And so, of course, your whole body presses down on the seat just a little less than before, since it weighs less. If you were rounding the curve fast enough, you really would be thrown off the seat entirely, freely falling into "space," as weightless as an astronaut; a bit of air resistance pushing back at you would be the only real distinction between you and an astronaut.

But what about the bottom of the track? Here your whole body wants to keep moving in a straight line, but the car wants to follow the track, and so you are pressed against the car by a force out of your control, and in that moment you weigh more than your usual weight.

This idea of increasing your weight by being held fast in a circular path is just the way that an astronaut, who otherwise would be floating about in space, could be made to feel right at home. Whirl an artificial home round and round its central axis–it could resemble a giant bicycle inner tube with the astronaut inside –and an artificial gravity is created.

To make him or her feel really at home, we could create the equivalent of the Earth's gravity, but we would need a rotational period measured in many minutes for a wheel of several kilometers' diameter.

The astronaut would quickly lose the feeling of Earthliness, however, as soon as he or she moves about, or moves things about. On the Earth, if we drop a stone from rest it falls straight down, or straight enough if we allow that the stone's motion is governed by the Earth's static gravity alone and not by the Earth's rotation. But release the stone in the Space Wheel and its motion appears curved–yet in the space of the Solar System it shoots out straight as an arrow, apart from local gravitational effects. The path, in nonrotating space, is a straight line tangent to the circle that it traced while held in the astronaut's hand. The astronaut's floor, the outer rim of the Wheel, intercepts the stone's short-lived trip through space.

It is rather important to remember that the artificial

gravity we create by rotation begins to give us problems as soon as we are no longer static. We stretch out our hand and it is pulled (or pushed if you like) off to the side. We walk toward the Wheel's axis and we veer in a curved path. And then when we stop, we discover that we have lost some weight.

The simulation of gravity is far from perfect, but it was never meant to be perfect. "Real" gravity, the pull of real masses toward each other, as we know, grows weaker as the distance between them grows larger. For point masses, or spherically symmetric masses, the attraction decreases with the square of the distance, that is, double the distance and the attractive force drops by a factor of 4. Halve the distance and the attraction quadruples. But if we should move toward the central axis of the Space Wheel, climbing "up" a ladder perhaps, the simulated gravity would now grow weaker and weaker. If we move halfway toward the axis, the "outward force," or simulated gravity, is only half as great, in sharp contrast to the way the gravity force changes with distance above a planet. (Bear in mind, however, that the gravity force beneath the Earth's surface grows weaker and weaker as we move toward the center. Similar to the restoring force at the end of a stretched spring, this force is *directly proportional* to the distance from the null point and is therefore called an "elastic force." The simulated gravity in the Space Wheel, however, while proportional to distance from the axis, is *not* elastic for it is directed *away* from the axis, not toward it.)

Fictitious forces

When we walk (or climb) toward the axis of the Space Wheel we are pushed (or pulled) to the side. When we then let go of our support we are pressed out toward the rim, our starting point. What is the nature of these forces?

We are so used to Newton's law of motion–force

equals mass times acceleration—that we sometimes forget that it has reality only in so-called inertial, or Galilean (that is, constant-velocity) reference frames. If we sit in a room and lay a ball on a table and it suddenly begins to move faster and faster, we tend to say, "Well now, there must be a force present causing it to accelerate, for it cannot accelerate out of nothing." It is so easy to forget that the room itself might be pulled along by some force, continuously accelerating, and that the ball might be the only thing in the room moving at constant velocity! Or, suppose we might lay the ball on the table in a room which is freely falling—

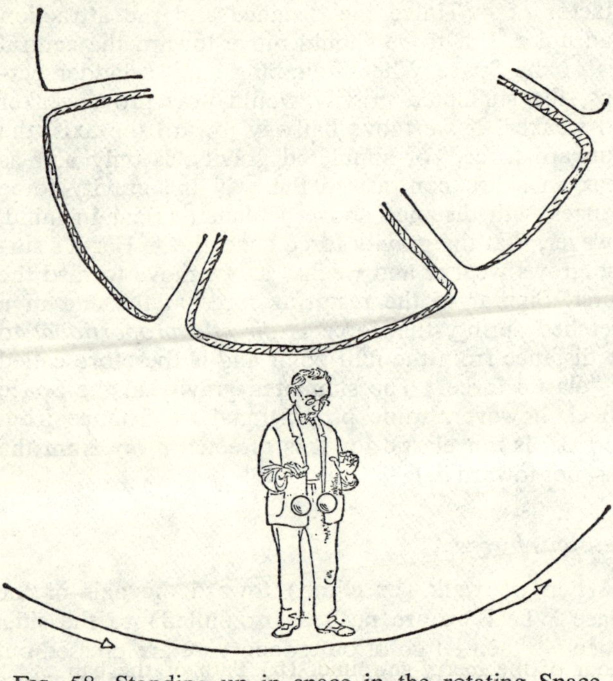

FIG. 58. Standing up in space in the rotating Space Wheel.

accelerating—in a gravity force field. If we lay the ball at rest, it remains at rest. If we give it a small velocity, the velocity remains unchanged. But all the while the ball, just as the room, moves under the force of gravity. And yet everything behaves within the room as it would far from any gravity field in interstellar or intergalactic space.

What is real? What is illusory? If we knew all the answers we would unravel many secrets of the Universe.

If you should throw a ball from a merry-go-round to someone standing outside, it would take great skill to reach him or her, for the ball would move in a curved path (in merry-go-round space), as if an invisible demon were pushing it to one side, but all that really happens is that your sideward tangential velocity, given to you by the merry-go-round, is added to the forward velocity of the ball sending it in a straight line that will never reach the receiver.

If we are uncomfortable (as we should be) with curved, apparently accelerated, motion taking place within the merry-go-round, yet with no force present,

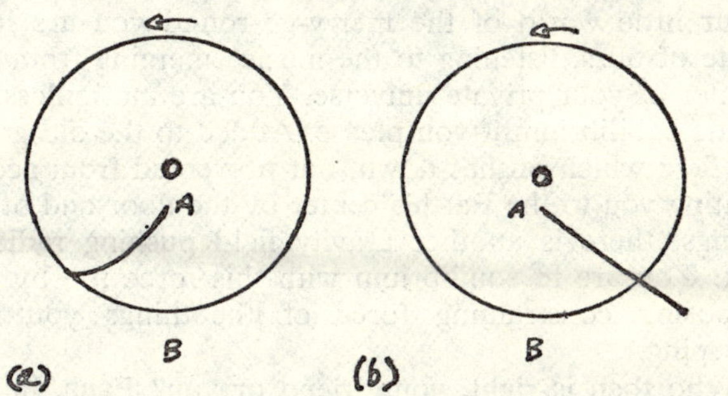

FIG. 59. A ball is thrown from a merry-go-round from *A* toward *B* outside. (a) Path the ball traces on the floor of the merry-go-round. (b) Path of the ball as seen by *B*.

then there is only one thing to do: Say that your invisible demon is really pushing with a force, and name it (the force, not the demon), Coriolis, after the man who first wrote about it. But of course it is no more real than the acceleration corresponding to it, for, as I have said, the ball "actually" moves in a straight line. You may have a bit of trouble convincing the fellow riding one of the horses of that fact, after he was struck in the head by the ball you were directing to your friend outside.

Real or fictitious, much depends upon the observer and his reference frame. As you are pushed against the edge of a merry-go-round the argument can be rather convincing that a very real force is centrifuging you. Your friend on the outside, knowing the facts, would certainly say: You are continuously changing your direction of motion as you hold on for dear life moving in a perfect circle. This means that wherever you are, you are accelerating toward the center. The constraining force—the horse, the rail, the floor—holds you in place and is also directed toward the center. How big is the force? This depends on your mass, for the acceleration is fixed by the turning rate and the radius. In your little world of the merry-go-round you are in a state of bliss, listening to the music emerging from the center of your private universe. You are motionless, in static equilibrium if you please. Added to the old gravity field which pushes down, but prevented from accelerating you to the Earth's center by the floor and other things, there is another gravity field pushing radially out. You are in equilibrium with this force too by the opposing constraining force of the things you are grasping.

Who then is right, your friend or you? Each, in his own way, is right. The gravity field demon does have reality, for even in a merry-go-round world the principle of equivalence must hold, at least in a small region of space. And when you walk from the center toward the rim, you do not merely *think* you are being pushed to the side, you *are* being pushed since your feet, with each step you take, strike portions of the floor *which are moving faster and faster through inertial space*. Reality again.

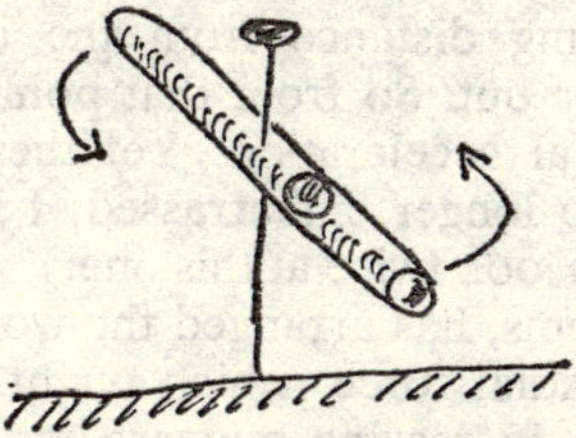

FIG. 60. The pea, constrained to lie within the rotating soda straw, will move out faster and faster–yet there is no radial force to propel it. Is Newton's law of motion violated?

A final point. By looking at an embarrassingly simple example, it might be easier to grasp some of these ideas. Imagine a pea placed in a soda straw, and the straw rotated about an axis, as in Figure 60, perpendicular to the axis of the cylinder. For simplicity rotate the straw at constant velocity. If we place the pea slightly off the axis of rotation, it will start to move out, faster and faster, until it leaves the straw. (This is a thought experiment; try it if you have doubts.) Question: Where is the force propelling the pea? In rotating space the answer is clear–centrifugal. It drives the pea out faster and faster. If you were to cork the end of the straw the radial acceleration would disappear, but not the centrifugal force which presses the pea against the cork which in turn presses back, just as gravity presses you against the chair which presses back as you read this. And even as the pea moves radially out, the Coriolis force, acting sideways, is held in check by the constraining force of the straw itself.

Now let us look at the pea in inertial space. Its path in space is clear enough–a *spiral.* But how could it have an increasing radial component of motion, a radial acceleration, without a radial force (repelling magnets? a spring?)? The answer is clear. It has *no* radial acceleration. Yet, with each tick of the clock it is at a rapidly increasing distance from the center, moving faster and faster out. So from that point of view there *must* be a radial acceleration. Yet there is *no* radial force. (I am no longer embarrassed; I have seen doctoral candidates look twice at this one.)

Nature, it seems, has arranged the world in a beautiful way. The radial force, which ought to be there to account for the increasing outward motion, is exactly canceled by the inward radial force (centripetal) that would account for the pea's motion *around* the center; so, at all times the radial force is *zero,* just as the *net* radial acceleration is zero.

I wonder how Aristotle would have attacked this simple problem. Isaac Asimov believes that if only the Solar System would have had a stellar companion with a planet or two of its own, all visible from the Earth with the naked eye, humankind might have bypassed the Ptolemaic geocentric world system and the tediously long Aristotelian era. Science would have advanced by hundreds of years.

Imagine, color TV in the fourteenth century!

CHAPTER SEVEN

The Force

Climbing into space

The town of Cocoa Beach, Florida, has the slogan: "The World's Finest Beach." Whether or not to believe it depends, naturally, on your own experience and requirements for the title. In a competition a contestant cannot also be judge! Some people may insist that a long row of motels, bars, and restaurants adds to the local color, especially if they happen to be sleepy, thirsty, or hungry. As for me, I think that the Great North Beach of a little island called Anholt is the world's finest, although I admit that I have never made a complete survey. Probably you have never even heard of Anholt. It is so tiny you would need a magnifying glass to find it on the map of Europe. You would have to look in the middle of a triangle formed by Jutland, Zealand, and Sweden. The beach curves like a crescent. Quite unlike Cocoa Beach, it would be hard to find from one end to the other a sign of human life anywhere, save for a solitary lighthouse at the east end. And the sand is pure white, hardened at the water's edge, so it is perfect for cycling.

But Cocoa Beach brings back memories of the Golden Sixties, where you can dig your toes in "talcum powder," forget the motels, and look up at the Saturns streaking by. At the north end of Cocoa Beach lie Cape Canaveral (named Cape Kennedy for a brief period) and the John F. Kennedy Space Center; at the south end is the Patrick Air Force Base. So, all through the sixties and seventies—through the Mercury, Gemini, and Apollo programs and the many satellite and planetary probe launches—Cocoa Beach was the western world's space capital.

I was fortunate in those days to be given the opportunity to teach concentrated course programs there in space science and technology. "AstroCourse," as it was familiarly called, tried to accomplish in six days what usually takes a semester at a university. We gradually came to realize that it was better to run it every second day, with one day in between, to give the students, and me, time to recuperate.

On one of the in-between days, while walking the length of Cocoa Beach with my feet half in and half out of the water, I thought about a problem I posed to the engineers and scientists of the class. Not that I

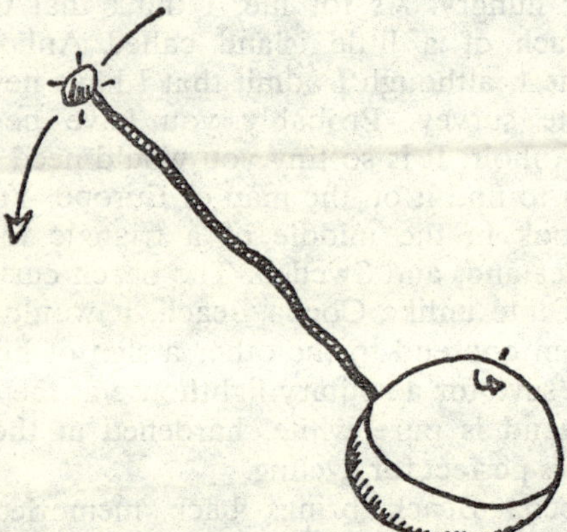

FIG. 61. Dropping a ladder from a satellite in *synchronous orbit* above the equator, that is, a satellite with an orbital period of 24 hours which will permanently remain above the same spot on Earth.

wanted to patent the concept, since we were all in it together; it was, how shall I say it, an idea waiting to be discovered.

Let us imagine a satellite in circular orbit about the Earth's equator, some 35,900 kilometers (22,300 miles) up. This is a *synchronous* satellite, the same satellite that appeared at the close of Chapter Two. It revolves about the Earth in 24 hours and so it is permanently fixed in the sky, as the "fixed stars" and Sun sweep past it. That means it is always hanging over the same point on Earth. Now suppose we let down a ladder. . . .

I can hear some of you saying, "Of course, a ladder from space to Earth! Why didn't anyone ever think of that before?"

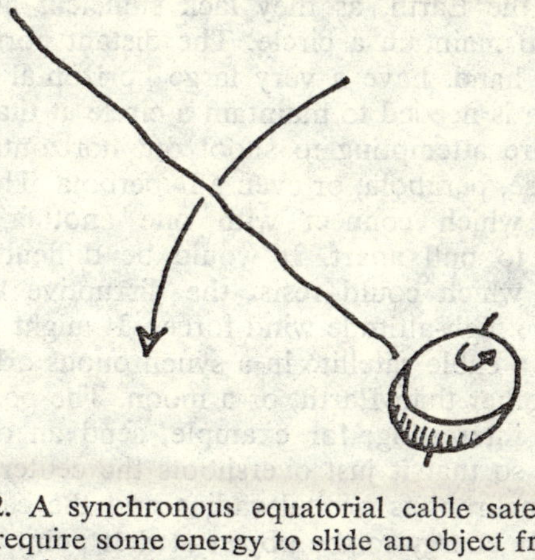

FIG. 62. A synchronous equatorial cable satellite. It would require some energy to slide an object from the Earth to the synchronous orbital height, but once it overshot that point the cable itself will furnish energy which will cast the object into space (see Figure 60 for a similar dynamical problem). Such a cable satellite would surely encounter problems of stability, i.e., its vertical orientation would not be restored if perturbing forces were to cause it to shift away from the vertical slightly.

Not even Jules Verne thought of it. Just consider the problems and then you will see why. Remember that for each radial distance there is a unique orbital period. Although it is 24 hours for that far-out satellite, it is a mere 90 minutes for a close-in one.

Even if you lowered a ladder, or cable, down to Earth and simultaneously raised another up to the sky, so that the center of mass of the orbiting satellite still remained at the 24-hour radius, the tension in the ladder or cable would be enormous. You would be requiring each element of mass to orbit at 24 hours even though it may be at a distance which corresponds to a circular orbit of 12 hours, for example, or 36 hours, or only 90 minutes. The portions near the Earth exert a pull on their neighboring members, for they are trying to fall right down in an ellipse which might intercept the Earth, as they lack sufficient horizontal velocity to maintain a circle. The distant portions, on the other hand, have a very large horizontal velocity, more than is needed to maintain a circle at that radius, so they are attempting to shoot out horizontally in a long ellipse, parabola, or even a hyperbola. Thus, mass elements which connect with one another have a tendency to pull apart. It would be difficult to find materials which could resist the disruptive forces as well as the high-altitude wind forces. It might be easier to design a cable satellite in a synchronous orbit about a planet other than Earth, or a moon. The possibilities might be interesting; for example, send an object up the cable so that it just overshoots the center of mass at the synchronous orbital radius and the cable itself will supply energy to the object at that point, injecting it into space, like the pea in the soda straw at the end of Chapter Six.

How Saturn got its rings

What about these disruptive forces trying to pull apart this strange cable satellite? We encountered them when we described the motion of the Earth and Moon about each other and their mutual tendency to bulge their bulk toward one another along their line of centers.

Remember, Newton pointed out that a spherical mass—or more properly, a spherically symmetric mass—behaves gravitationally just as if it were a single mass point located at the geometric center. Whether the sphere is a giant ball of iron, or a ball of sand, or dust, makes no difference. Place the sphere in space and the total gravitational force on it is directed from its center, as if its total mass were right there.

But what about the effect of the inverse-square gravitational force field on it, say, from the Earth, pulling harder on the center portions than the more distant portions, and much harder on the close-in portions compared with the center? The masses will be stressed by different amounts because of these differing forces, and therefore will try to accelerate from each other. But will they be able to accelerate?

If the satellite were a ball of iron, tensile forces will hold it together; it will continue to move approximately as a sphere, though slightly strained by these "tidal stresses."

But if the sphere were a ball of sand, or dust, or liquid, not even the cohesive force of mutual gravity among the mass elements would be sufficient to prevent tidal disruption, unless . . .

Unless the ball of sand, dust, or liquid were a far-out satellite, so far that there would not be much difference between the gravity force from the Earth on the close-in mass elements and on the more distant elements. Or at least the difference would not be as great as the attraction of the elements toward one another. Obviously there must be some critical distance above which such a loosely-coupled mass system will continue to be held together as a satellite and below which it will disintegrate, with each mass element becoming an independent satellite.

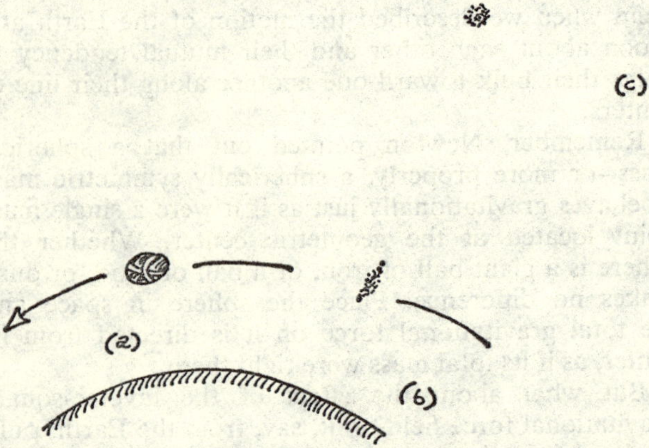

FIG. 63. (a) A satellite held together by nuts and bolts or by other solid-state forces will, for the most part, continue to hold its shape even though it is stressed to bulge toward and away from the Earth.

(b) A sphere of sand in a low orbit, on the other hand, cannot hold together. Each grain will accelerate by a different amount depending on its distance from the Earth. The grains will move in separate orbits and gradually the sand will encircle the Earth.

(c) If the spherical mass of sand were placed in circular orbit of radius at least 2.44 times the Earth's radius, and the sand and Earth were of the same density, the sphere would be able to retain its shape. Then on the average the difference in the Earth's force of gravity from one grain to another would not be as great as the gravitational attraction of one grain toward another.

In the early years of the Space Age the United States planned to release a sackful of needles in orbit and let them gradually form a ring about the Earth. Each needle was, so to speak, a satellite on its own, and in its own orbit, acting as a small dipole, part of a giant world-wide reflector for microwave communications. You can imagine the anger of radio astronomers and others crying "foul!"

This marked the beginning of greater public concern for the global environment. There are examples enough in recent years of serious problems which extend beyond the borders and the decision-making of any one

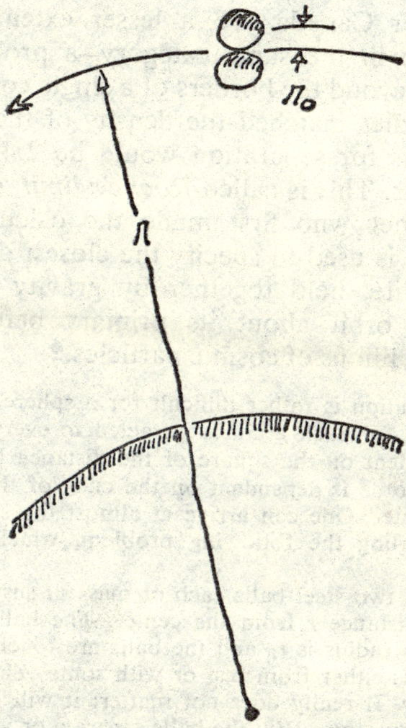

FIG. 64. Two balls above the Earth, suspended at rest or in orbit. They will tend to fall at unequal rates because of the unequal force of the Earth's gravity on one relative to the other. This tendency increases the closer they are to the Earth, and they will separate unless the mutual force of gravity between them were stronger than the difference noted above, as would be the case at a sufficiently high altitude. If the density of the balls were the same as that of Earth, the critical distance from the Earth's center above which separation would take place would be *twice* the Earth's radius.

nation. A supertanker breaks up off the Brittany coast and discharges its oil cargo on the shores of three countries. Coal and oil power plant emissions from all over Europe fall down on Scandinavia in the form of sulphuric acid. A radioactive satellite falls down unexpectedly over Canada. To a lesser extent the space needles were in the same category—a problem which extends far beyond the borders of a single country.

If the needles matched the density of the Earth, the critical radius for separation would be 2.44 times the Earth's radius. This is called *Roche's limit,* named after the astronomer who first made the calculation. The word "limit" is used to specify the closest distance that a protosatellite, held together by gravity in a liquid state, could orbit about its primary before disintegrating into billions of cosmic particles.*

* The calculation is rather difficult for a sphere composed of grains of sand, for each grain is attracted to every other in an amount dependent on the square of the distance between them and the tidal force is dependent on the cube of the distance to the Earth's center. One can arrive at almost the same result as Roche by tackling the following problem, which is infinitely simpler:

Let us imagine two steel balls each of mass m suspended above the Earth at distance r from the center. The balls are aligned vertically; each radius is r_0 and the balls are touching. Now we release the balls either from rest or with some velocity, even an orbital velocity. It really does not matter; it will not affect the answer to the question: Will the balls separate or will they move as one?

We know that the force of separation between one of the balls and the center of mass of the two is proportional to

$$\frac{m_e m}{r^2} - \frac{m_e m}{(r + r_0)^2}.$$

The mass of the Earth, which defines the strength of the gravity force field at r or at the point at the center of the upper ball, is m_e. This force difference can also be written as

$$\frac{2m_e m r_0}{r^3},$$

provided we assume that r_0 is very much smaller than r, which it is of course especially if we can hold the balls in our hand.

Which brings us to the case of Saturn. No one really knows, of course, how Saturn got its rings, but one guess is that a satellite simply came too close and disintegrated. This is a poor guess, it turns out, because the rings are too far in, from 1.3 to 2.3 planetary radii. One might also postulate that an external body, an asteroid for instance, might have come close enough to be pulled to pieces by the gravitational tidal forces from Saturn, or the asteroid might pull out portions of the liquid/gaseous protoplanet streaming toward it. But the approach speed would be hyperbolic or parabolic. What mechanism could account for particles that move in a circle? And why such small particles (to be consistent with the extreme thinness of the rings)? The most logical theory seems to be that as the primordial gas and dust in the region of the outer planets condensed to form planets and satellites, one planet in particular, Saturn, had a larger and longer lasting envelope of gas. This gas produced a drag force which plunged orbiting particles into the planet. Water vapor, a less volatile

---

The mutual force of gravity directed toward their center is dependent on $2r_0$, the distance between their centers:

$$\frac{m^2}{(2r_0)^2}.$$

We now can find the *critical* distance from the Earth's center, above which the balls hold together and below which they will go their separate ways. All we need do is equate the expression for the force of separation, the force difference, and the force of gravitational attraction. Before we do this, let us make a convenient assumption—the density of the balls is the same as the Earth. This means that the mass either of a ball or the Earth itself is proportional to the cube of their respective radii, for mass is equal to density times volume. We immediately conclude (try it): The critical orbital radius is *twice* the radius of Earth. It is a bit less than 20% under Roche's limit. Apparently a ball of sand will not hold itself together as well as the two contiguous balls, for the sand ball will begin to disintegrate at the higher altitude. Note that the radii of the balls play no part in either calculation, so long as we do not tamper with the basic assumption that the balls are much smaller than the Earth.

gas than most, condensed into chunks of ice not very much larger than icebergs. They too experienced drag in the now thinning gaseous corona, the effect of which, together with their mutual collisions, was to

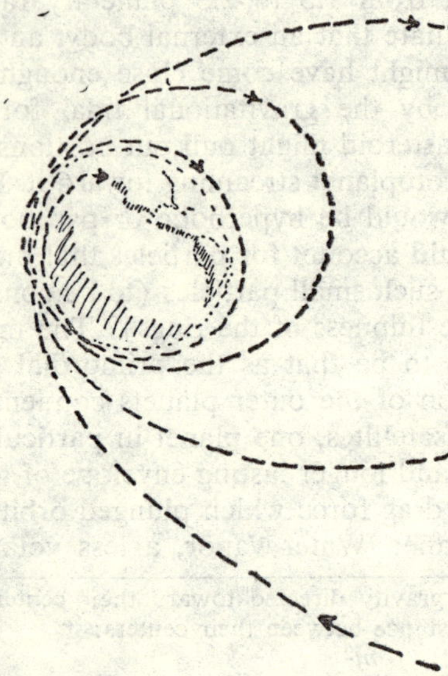

FIG. 65. How to remove orbital energy without using up propellant. A satellite enters the upper atmosphere in a long ellipse, almost parabolic, and receives a strong drag force at perigee. Such a satellite could, for example, be returning from a lunar mission. The heat developed is radiated away into space in the new shortened orbit. On re-entering the atmosphere, it again receives a strong drag force. As orbital energy is again radiated as heat, the ellipse once more grows shorter. Eventually the orbit is "circularized" and atmospheric drag takes place all along the orbit, as described in Chapter Three. The loss in orbital energy is then seen as a drop in altitude in the converging spiral, or severe loss of potential energy, even though the satellite speeds up and increases kinetic energy.

gradually change their original elliptic orbits into circles. This circularizing of orbits by means of a gaseous *perturbative drag* force is a well-known practice in orbital mechanics. A satellite enters the Earth's upper atmosphere at an almost parabolic speed and at each pass at perigee it loses some orbital energy; the ellipse grows shorter and shorter. In the case of the orbiting icebergs of Saturn the gaseous envelope would eventually have to vanish, otherwise the drag would appear at all parts of the circle and there would be hailstorms of almost iceberg size on Saturn as frequently as rain showers on the Earth.

This same reasoning might have accounted for a ring system about Jupiter, but because of its larger mass, by a factor of 3, it was far more luminous than Saturn at the time of satellitic formation and too hot to allow water vapor to condense into ice. It turns out to have a very fine ring. Uranus (and perhaps Neptune), on the other hand, also appears to have a barely visible ring system.

The rings of Saturn were discovered by Galileo in 1609 soon after his new telescope revealed the four largest moons of Jupiter (the Galilean Moons). But Galileo was not sure of what he was observing. Perhaps they were bright satellites on either side of the planet, yet they disappeared and reappeared in the same place. It was not until 1655 that these strange appendages to the planet were identified, by the young Christiaan Huygens, as a single object, a thin flat ring about the Saturnian equator. Twenty years later Cassini discovered the dark gap, named after him, which separated the ring into two annular parts. Seventy-five years after that the third ring, the "crepe" ring, was found closer to the planet's surface.

To anyone unfamiliar with the elementary principles outlined in the preceding section the immediate thought might be that we have here three annular solid disks turning round, similar to the outer sections of a phonograph record. Even the great eighteenth-century mathematician, Laplace, maintained that the rings were composed of continuous matter, or at least of solid ringlets moving within one another. James Clerk Maxwell, very likely the most renowned mathematical physicist of the nineteenth century, set his mind on the ring problem and proved that they must be composed of independent tiny satellites. As noted above their numerous colli-

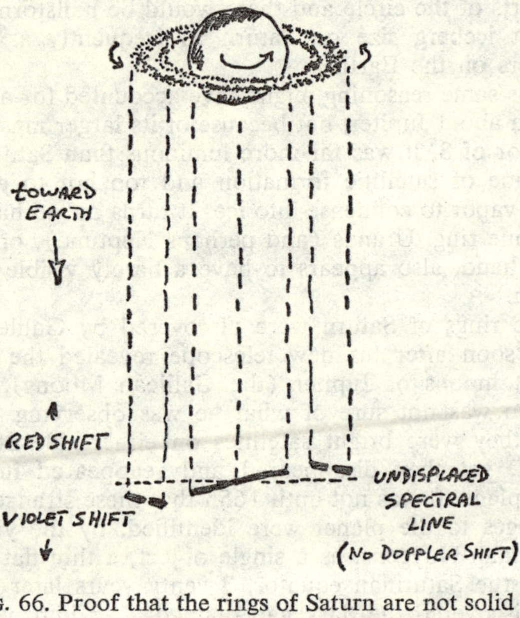

FIG. 66. Proof that the rings of Saturn are not solid flat rotating masses. If they were solid, the outer portions would move through space faster than the inner ones and this speed would be revealed by a *Doppler shift* of the spectral lines of solar radiation reflected by the rings as viewed with spectroscopes on the Earth. But just the reverse occurs. The outer portions move more slowly through space, revealing that the rings are composed of independently orbiting masses.

sions tend to circularize the orbits; they also flatten the plane to a thickness of several kilometers at most.

But what of the gaps between the rings? These can be explained by a gravitational pull, acting as a perturbative action, by the satellites of Saturn, disturbing the controlling force of the planet upon the ring particles. A particle travelling in Cassini's division would have an orbital period exactly half of that of Mimas, so that distances of closest approach to Mimas will keep recurring. This amplifies the perturbative disturbance and eventually sweeps the gap clean of ring particles.

It was not long before the spectroscope was used to verify Maxwell's conclusion. We know from the work of Christian Doppler, that waves, light or sound, from an approaching source or reflector will shift in frequency and wavelength in an amount depending on the speed relative to the receiver. If the speed is negative (i.e., if the source or reflector were receding), the shift in frequency and wavelength will be in the opposite direction. And so in 1895, James Keeler of Allegheny University used this principle of Doppler, almost for the first time, to demonstrate that the ring particles really do move as independent satellites, by noting the shift of solar absorption lines reflected by approaching or receding particles. Ring particles which were moving toward Keeler's telescope produced a spectral shift toward the violet end of the spectrum. The shift would be less for those particles orbiting in the outer parts of the ring for they move more slowly through space; if a ring were an annular disk, on the other hand, the outer portions, moving faster, would have shown an even more pronounced Doppler shift than the inner portions.

But what has all this to do with space flight? Tidal effects are only one of the many effects, although a rather important one, which can influence the orbital motion and orientation in space of a space vehicle. A "third-body perturbation" such as that of Mimas on an orbiting ring particle, or of Mars on a spacecraft headed for Jupiter, is of immense interest and significance in space dynamics, as we will see in the next section.

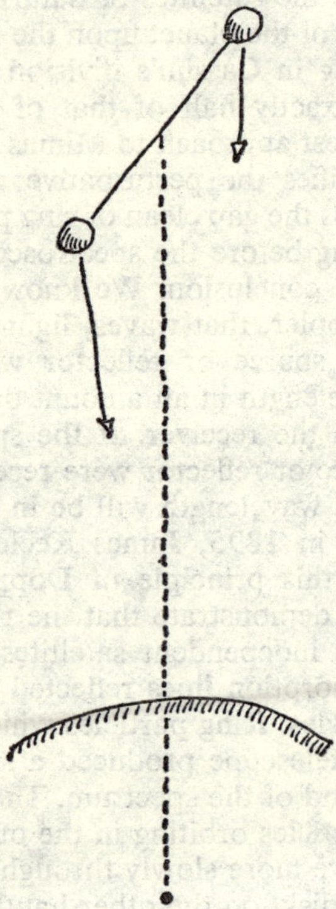

FIG. 67. All satellites which are not spherically symmetric tend to oscillate like this dumbbell in orbit. The lower knob receives a stronger gravitational pull than the upper one, and the net effect is an oscillation about the vertical orientation. There are no damping forces, such as air friction, which would have been able to dissipate the mechanical oscillation energy as heat and bring the satellite eventually to rest in the vertical alignment.

As for tidal forces, the general rule is: The larger an object and the lower its orbit the more it will be influenced by these forces of separation. As we know, the gravity field of a point source or spherically symmetric source weakens with the square of the distance from the point or center of the sphere. The tidal force is much more sensitive to location; it weakens with the *cube* of the distance. A dumbbell satellite is a case in point. Suppose it swings about the Earth in an orbit low enough so that the difference between the forces on the widely separated balls will be significant. Unless the dumbbell lies exactly horizontally or exactly vertically, this force difference will tend to swing the satellite back and forth like a pendulum about the center. The neutral orientation about which the swinging takes place is the vertical one. It tries to line up vertically but it never really can. This is because the swinging motion causes the dumbbell to overtake this neutral orientation since there is no air friction to dampen the motion and bring it gradually to rest. The period of the oscillation is about 60% of the orbital period, or just under an hour for a low-orbiting Earth satellite. It is difficult to design a satellite which does not have some characteristics of a dumbbell, even if it does not resemble one. Imagine the problems of stabilizing the radio astronomy satellite with crossed antennae each as long as the Sears Tower. Or the orbiting solar power collector, stretching over the area of four football fields. It is almost imperative that they are lifted far above the Earth, if only to reduce these tidal oscillations.

When does a perturbative force stop being a drag?

As the name implies, and in the sense in which I have been using it, a perturbative force is a slight push or pull on an object in space apart from the main force controlling its motion. For example, if the Earth or an interplanetary spacecraft orbits about the Sun, the rela-

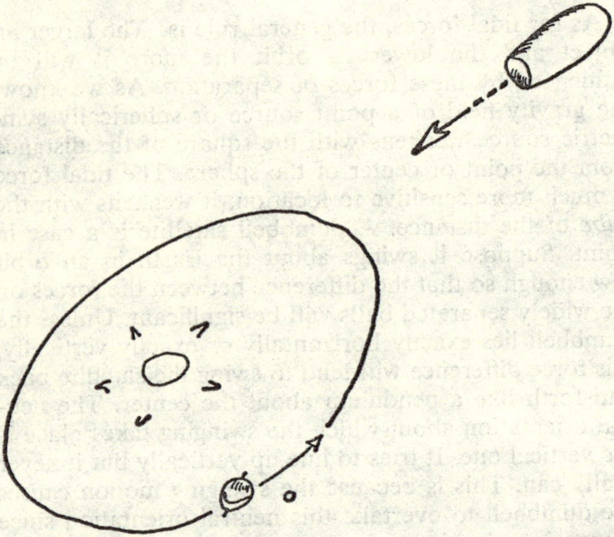

FIG. 68. A massive object that entered the Solar System and then departed could cause a cataclysm. It need not emit intense radiation nor even come close to the Earth to destroy its inhabitants. All that is needed is that the Earth's solar ellipse be perturbed by just enough to upset the climate and the delicate balance of nature which depends on it.

tively tiny forces of the other planets on them will influence their motion enough to bend their orbits out of their "pure" ellipses. In *Rendezvous with Rama,* Arthur C. Clarke speculates on the effect of an intruder in the Solar System, the entrance of a foreign massive body, even a small one, a collapsed dead sun perhaps, "a madly spinning sphere of neutronium, every cubic centimeter weighing billions of tons. . . ." It would "completely distort the orbits of the planets. The Earth had only to move a few million kilometers sunwards—or starwards—for the delicate balance of climate to be destroyed. The Antarctic icecap could melt and flood

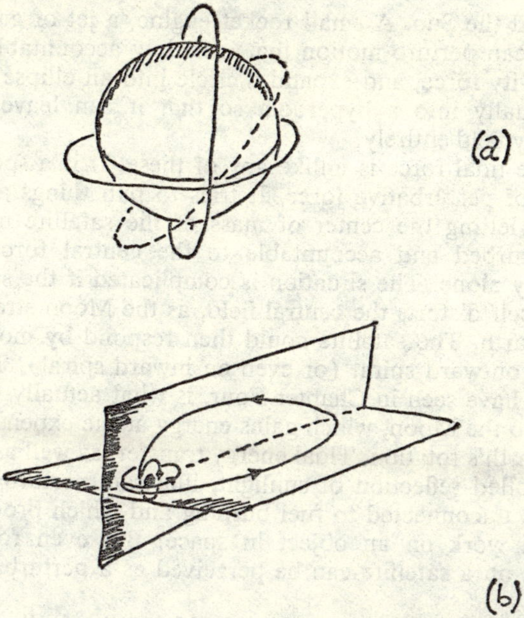

FIG. 69. (a) A perturbative rocket thrust, applied each time around at the same point in the orbit, can turn an equatorial satellite into a transpolar one.

(b) This same end result can be accomplished with 40% less propellant by first thrusting the satellite into a long equatorial ellipse; then, at apogee, where the satellite is moving at negligible speed, the orbit can be turned into a transpolar ellipse by a diminutive thrust. At perigee, the ellipse is then circularized by applying a reverse thrust.

all low-lying land; or the oceans could freeze and the whole world be locked in an eternal winter. Just a nudge in either direction would be enough."

Sunlight is also a perturbative force. We have already seen how it can pull a spacecraft out of its ellipse and spiral it out of its gravitational hold, even spiral it toward the Sun. A small rocket engine, a jet of gas or ions, can perturb motion that is mainly accountable to a gravity force, and expand a circle into an ellipse and eventually into a hyperbola so that it can leave the gravity field entirely.

The tidal force is unlike any of these. It is a special kind of perturbative force. It tries to pull things apart while letting the center of mass of the satellite move undisturbed and accountable to the central force of gravity alone. The situation is complicated if the satellite itself distorts the central field, as the Moon stresses the Earth. The satellite could then respond by moving in an outward spiral (or even an inward spiral). This, as we have seen in Chapter Four, is what actually happens to the Moon, which gains energy at the expense of the Earth's rotation. Tidal energy transfer, as well as the controlled reflection of sunlight, illustrate perturbative forces unconnected to fuel burning and which produce useful work on an object in space. But even rocket thrust on a satellite can be perceived as a perturbative force.

The most ingenious and valuable case of all, provided entirely by nature, is the third-body perturbative energy transfer known as the *swingby*. To describe it, it is best to review the facts behind the two-body problem.

The two-body problem is, as we know, the *Keplerian* problem. A satellite racing around the Earth is one example. Another is the Earth orbiting about the Sun. Then why is a satellite rounding the Earth which is veering about the Sun *not* a three-body problem? It is, of course, only it is far easier to analyze it as two two-body problems, as we have seen earlier, each in the presence of a perturbative force. The satellite orbits in an ellipse, but the Sun pulls harder when the satellite is near the daylight side of the Earth than when it is on the night side. And so the Earth-dominated ellipse is distorted by the Sun. The Earth and satellite move as one about the Sun. But if the satellite is massive, as is the Moon for example, the mass center of the combination shifts toward the Moon and the resulting path that the Earth follows about the Sun is a rather wavy ellipse. The Sun, of course, orbits about the galactic mass, carrying its planetary system with it.

If the Earth satellite is not a satellite at all but an interplanetary meteor which approaches the Earth at high speed, things can get complicated. Close to the Earth the situation is classically Keplerian, a more or less pure hyperbola. But as the meteor recedes from the Earth, the Earth's gravity gets weaker and weaker as the long arm of the Earth-centered hyperbola merges with its great Keplerian ellipse about the Sun. The Earth, once the dominant influence, now becomes a minor disturbing source of gravitational force which affects only slightly the solar-centered ellipse. The detailed motion during the transition from an Earth-dominating influence to an Earth-perturbing influence is somewhat difficult to analyze, but the space mechanician is aided first by the inability of the meteor to influence the motion of the Earth or the Sun, and second by the ability of high-speed computers to converge readily on a solution to the accuracy required. This is a subject that we will look at further.

But for now, let us see how the swingby perturbation actually works. Imagine an interplanetary spacecraft approaching a planet hyperbolically, as a meteor would. It is hard to imagine how it could approach in any other way. Surely it cannot expect to enter at more than zero speed the planet's *sphere of influence*—which we will arbitrarily define as a sphere where the planet's gravity drops to 1% of its surface value, ten radii out, barely observable on a Solar System scale—and expect to orbit in an ellipse about it. Its speed in the *planetocentered* reference frame, when it is far beyond the sphere of influence—that is, the difference in speed between spacecraft and planet in the reference frame of

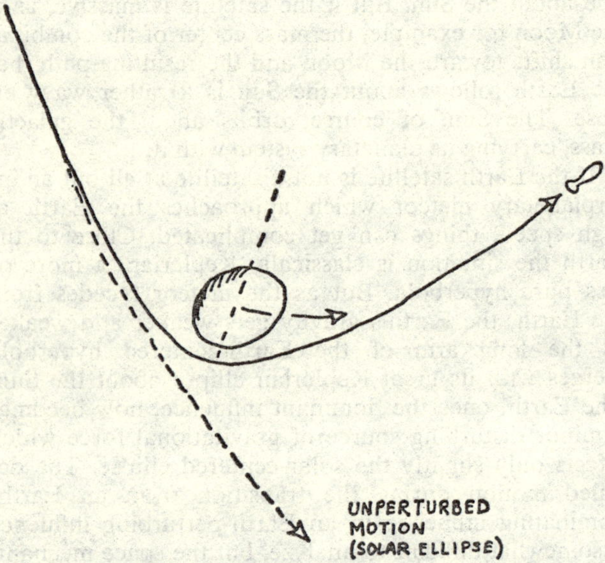

FIG. 70. A hyperbolic swingby. A hyperbolic encounter with an approaching object can result in a change in both speed and direction of that object.

the Solar System—is called the *excess hyperbolic speed* of the approaching (or receding) vehicle. To put it another way, in the sky of the planet a vehicle will approach from "outside," i.e., beyond the effective range of its gravity field, with this finite velocity. Energy must be conserved in this planetary reference frame as in all reference frames; thus, the vehicle "falls" faster and faster as it moves in closer and closer, that is, it moves down into lower and lower potential energy regions as the gravity field gets stronger and stronger. But unless it is heading onto a planet's surface it will reach a point of closest approach. We have called this point perigee for the Earth; it is simply the *pericenter* for any gravitational field source. At this minimum point it is not only moving horizontally in the planet's sky but moving most rapidly about the planet. As it continues to move it lifts further and further away and completes the open curve, the hyperbola, which is symmetric about a line connecting the planet's center with the pericenter. The speed of recession at every instant is identical with the speed of approach at equal distances from the planet. And finally, as it fades toward "infinity" it once again moves at the excess hyperbolic speed in the region of space where the planetary gravity field is negligible.

Now let us suppose the planet to be moving about the Sun *in the same general direction* as the spacecraft. As the spacecraft approaches and leaves the planet with the *same* speed and kinetic energy *relative to the planet,* then, measured in the frame of reference of the Solar System, it must have gained some small fraction of the planet's kinetic energy. It is the same as a *collision* between two objects which never actually touch, such as a magnet which pushes another with like polarity about through space. The planet actually imparts a bit of its orbital motion in the hyperbolic encounter, too small, of course, to be perceived by observing the planet, yet of inestimable value to the vehicle. As an added by-product of the swingby, the vehicle's *direction of motion* can be changed, if desired—without expending energy—and this, to be sure, is put to great use in many types of space missions.

There are a number of interesting situations in which the swingby maneuver is valuable. For example, a planetary satellite may play a role in aiding a spacecraft's escape from the planet's gravity field. It may impart some energy as well as help place the spacecraft "on course." The giant planets, Saturn and Jupiter, may contribute sufficient energy and direction to an approaching interplanetary probe to enable it to move further on toward the outer planets and beyond, exactly as has occurred in the case of Pioneers 10 and 11.

But of all the swingby cases none is more interesting than the excursion trip to Mars by way of Venus. Now the usual trip to Mars is taken by the familiar cotangent ellipse (also called the Hohmann transfer orbit

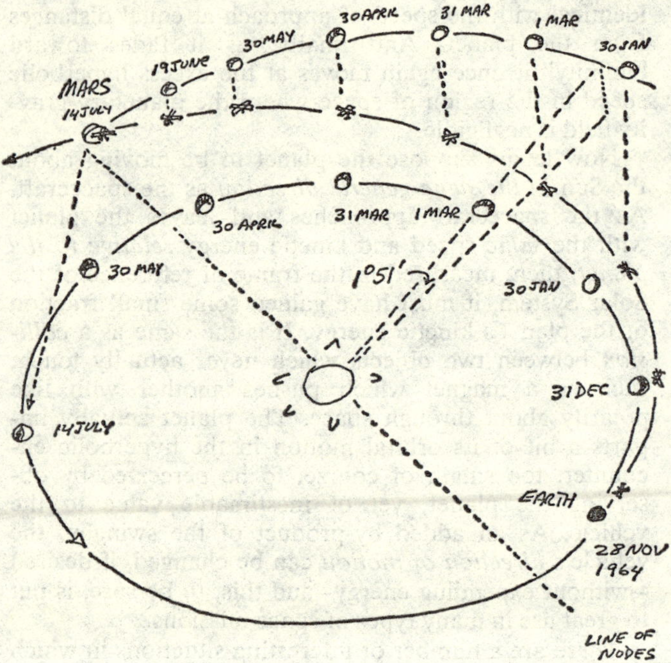

FIG. 71. A typical voyage to Mars by means of a cotangent, or Hohmann, semiellipse. This is the historic trip of Mariner IV. Note that the contact point was planned to occur when Mars was to enter the Earth's orbital plane. This minimized the propellant inventory by permitting the spacecraft to orbit completely in the Earth's plane. The line of nodes is the intersection of the two planetary planes inclined at 1°51′.

from the German space pioneer who first described it). This trip, as we have seen in Chapter Three, is strictly economy class expending the minimum propellant possible despite what appears to be a roundabout journey. The trip takes seven and a half months. Nevertheless, if an expedition were sent to Mars it would have to wait

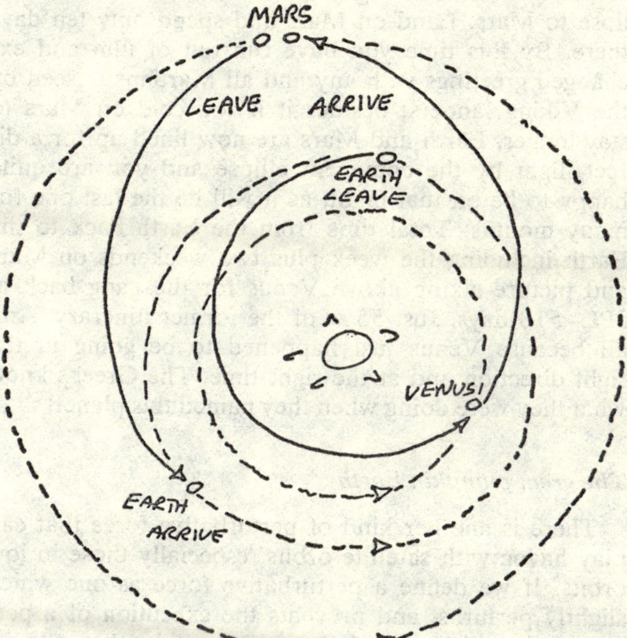

FIG. 72. The "Venus swingby" voyage to Mars. A hyperbolic encounter with Venus transfers energy to the spacecraft and also alters its direction so that it is able to move out to the Martian orbit. This brings the Earth to just the right position to meet the spacecraft returning on an ellipse after spending ten days on Mars. Without the swingby maneuver the round trip to Mars would take more than a year longer.

around for a year and four months before Earth and Mars are again lined up to make the return trip in a similar manner. Total journey: over two and a half years.

Is this really the best way to travel to Mars? Consider the following alternative itinerary: Interplanetary ellipse *to Venus*. Swingby Venus to produce a new ellipse to Mars. Land on Mars and spend only ten days there. By this time you have run out of film and exchanged greetings with any and all Martians unseen by the Viking landers; besides it is too cold on Mars to stay longer. Earth and Mars are now lined up for a direct flight by the cotangent ellipse and you are quite happy to be on that flight, as it will be the last one for many months. Total time from the Earth back to the Earth including the week plus two weekends on Mars and picture-taking above Venus for the gang back at JPL–510 days. Just 55% of the former itinerary. And all because Venus just happened to be going in the right direction and at the right time. The Greeks knew what they were doing when they named this planet!

The great pumpkin Earth

There is another kind of perturbative force that can play havoc with satellite orbits, especially those in low orbits. If we define a perturbative force as one which slightly perturbs, and prevents the execution of a perfect circle, ellipse, parabola, or hyperbola, then our list grows larger. We have already noted some of them: interfering "third bodies," air drag, electromagnetic radiation, interaction with magnetic, or possibly electric, force fields in space, and bombardment with charged particles and meteor streams. And then of course there is that special item called tidal force which makes itself known in two ways: 1) It interacts differently with the atoms that extend across the orbiting object, thus turning the object about its center of mass, and 2) it disturbs the orbit itself, for this object, if it were massive enough, presents a field across the extended primary source. This field may distort the shape of the primary so that it is no longer capable of maintaining Keplerian motion.

The Earth–Moon system is a classical example of these two effects. The Moon's *triaxial ellipsoidal* shape causes it to rock back and forth. This effect, called *libration,* corresponds to the oscillations of a dumbbell satellite which we encountered in the last section. Another libration occurs because the Moon executes an ellipse rather than a circle about the Earth. This means that its angular speed of *revolution* is greatest at perigee and least at apogee. On the other hand, its angular speed of *rotation* about its own axis is constant at all times, and equal to the *average* of the *revolutional* angular speed about the Earth in orbit. If the rotational and average revolutional angular speeds (or the respective periods) were dissimilar, the Moon would display all sides to us. The perigee–apogee departure from average, the ellipsoidal oscillation, as well as the different views one can get from widely separated observing points on Earth, enable us to see not 50% but 59% of the lunar landscape from our home planet.

The Moon's effect on the water mass of the Earth accounts for most of the Earth's tidal bulge, which continuously acts to destroy the lunar elliptical orbit. Recalling Chapter Four, the net effect is a spiraling of the lunar trajectory away from the Earth. Needless to say, a planet rotating very slowly compared with the revolution of a satellite would produce the reverse effect: The planet's own tidal bulge toward its satellite will lag behind it and the result will be a *drag* force on the satellite. The satellite will spiral in toward the planet, a fate destined for the Moon itself many billions of years hence. This investigation into tidal interactions provides a good answer to the question: Why do we find no satellites around Mercury, Venus, and (possibly) Pluto?

The *permanent* distortion of the Earth's shape, caused by its rapid rotation in its distant past when the Earth was cooling and hardening, and the response of its present plastic shape to its continuous rotation, accounts for a figure that is flattened at its poles and bulging at its equator. Such a figure, called an *oblate spheroid,* pulls a satellite orbit out of its plane so that

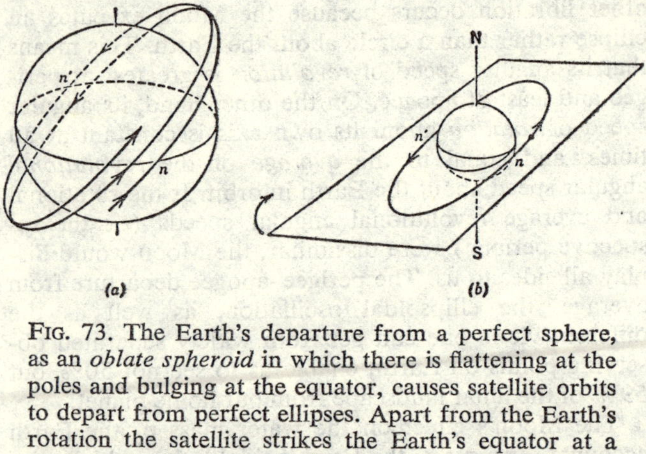

FIG. 73. The Earth's departure from a perfect sphere, as an *oblate spheroid* in which there is flattening at the poles and bulging at the equator, causes satellite orbits to depart from perfect ellipses. Apart from the Earth's rotation the satellite strikes the Earth's equator at a more westerly point each time around, as shown in (a). This can be interpreted, as in (b), as a *retrograde motion,* about the polar axis, of the plane containing the orbit. The plane moves in the direction shown by the arrow; it maintains a constant angle with the axis. When this angle is less than 26.5°, that is, when the satellite is almost transpolar, the major axis of the ellipse turns in its plane also in a retrograde motion; when greater than 26.5° it turns in a forward motion. Early satellite orbital studies confirmed that the Earth departs also from the oblate spheroid; it is, in fact, very slightly pear-shaped.

strictly speaking it no longer can be referred to as an orbit, a closed curve; however, it is convenient to think of it as a planar figure, an ellipse, where the plane itself steadily rotates about the Earth's north–south axis. Furthermore, the ellipse itself is caused to rotate within the plane. If the orbital plane is arranged so that it is inclined 63.5° to the equator, this rotation of the ellipse's major axis fails to take place. For this reason many of the early satellites have been launched at this angle just to keep the perigee point close to the launch area each time around. The period, either of the orbital plane's rotation or that of the ellipse within the plane, is always very long compared with the orbital period itself. This identifies the distorted portion of the Earth's field caused by the equatorial bulge as a true perturbative force.

The three-body problem according to Lagrange

The two-body problem, or Kepler problem, has such a neat and simple solution, a conic, that just about everyone (meaning students of elementary mechanics) used to think that the three-body problem—the analysis of the separate motions of three objects placed at random and with random motion, subject only to their pure point-source gravity fields—might well be 50% more difficult. It turns out to be (literally) infinitely more difficult to determine, for example, where all three are going to be, and how fast, and moving in what direction x seconds from now in the future or back in the past. This was very distressing and would have long ago shut down Goddard, Kennedy, and the Jet Propulsion Laboratory were it not for three factors which saved the day: 1) Most multibody problems are handled as a two-body problem disturbed slightly by other gravitational sources. 2) The *restrictive* three-body problem—where one of the bodies, an interplanetary spacecraft for example, is unlikely to pull the

Earth or Sun off its course—is relatively easy to solve. 3) Electronic computers, changing from analytical to numerical solutions, were invented in the nick of time.

In 1772, the great French mathematician, Joseph-Louis Lagrange, found an interesting special solution to the three-body problem. He realized that three bodies can move "in *constellation* with one another," that is, maintain their relative distances, when they are at the corners of an equilateral triangle. Specifically, if a third body enters a region of space at just the right velocity, it will be permanently fixed with the others at the ends of an equilateral triangle, one which may continuously vary its size. No matter. What is important is that the three move about together and that these apexes are regions of *stability*—if one wanders away unbalanced forces will appear which push it back. What perfect places for either an orbiting telescope or a space colony! Imagine either one of the two places in the plane of the Earth–Moon system that is at the corners of the equilateral triangles formed with Earth–Moon as base; the spot is bathed in almost perpetual sunshine and the temperature is maintained constant. But more about that later. We can convince ourselves that Lagrange was correct by pointing our telescopes in the regions of space which are in the orbit of Jupiter, but as far from Jupiter as Jupiter is from the Sun. There we will find groups of asteroids wandering about, but unable to wander away.

To keep the record straight, there are three more Lagrangian points. These are the *straight-line* solutions. The logic is simple and straightforward: Suppose a satellite orbits about a planet, and another satellite finds itself in a straight line on the outside. This second satellite thinks it is in a stronger-than-usual gravity field and if it is in just the right place it can go around in its ellipse in exactly the same time as the closer-in satellite even though it is further away from its primary. The orbital period, you recall, depends on dis-

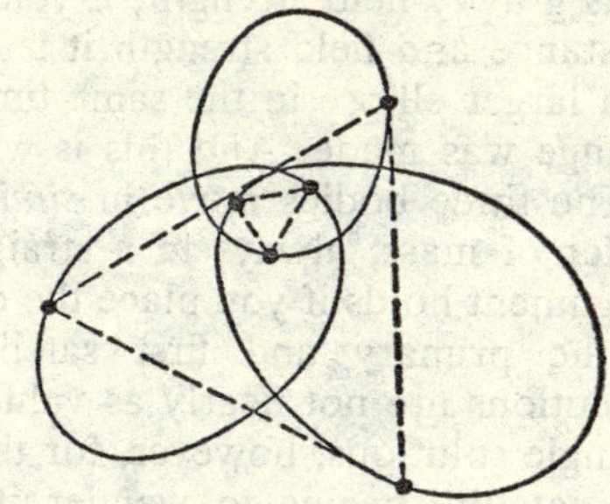

FIG. 74. A third body placed with the right velocity at one of the two "equilateral-triangle Lagrangian points" of two bodies in orbit will always keep equidistant from each of the two by an amount equal to their distance of separation. If it wanders away, gravity forces will push it back giving evidence of stability.

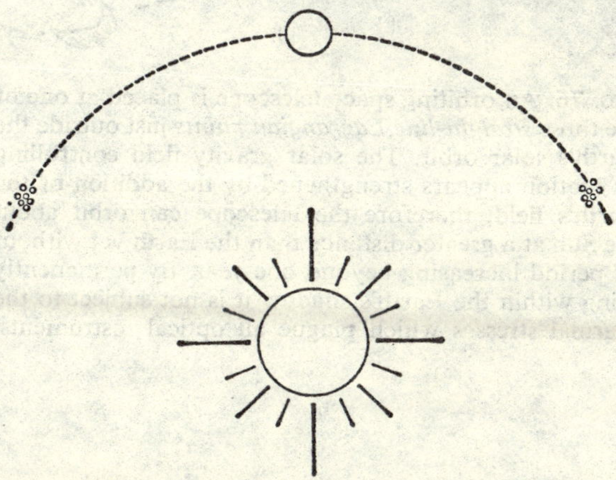

FIG. 75. The Lagrangian satellites, captured asteroids, of Jupiter.

tance as well as gravity field strength; if you should increase *both* distance and field strength it is possible to go around in a larger ellipse in the same time as it did before the change was made. And this is what is happening here. The three bodies move in *similar* ellipses about the center of mass, always in a straight line. A comparable argument holds if you place the outer satellite between the primary and first satellite. These straight-line solutions are not nearly as valuable as the equilateral-triangle solutions, however, for they are unstable—once a satellite begins to wander it keeps on wandering. Just as anything would that has lost the Force.

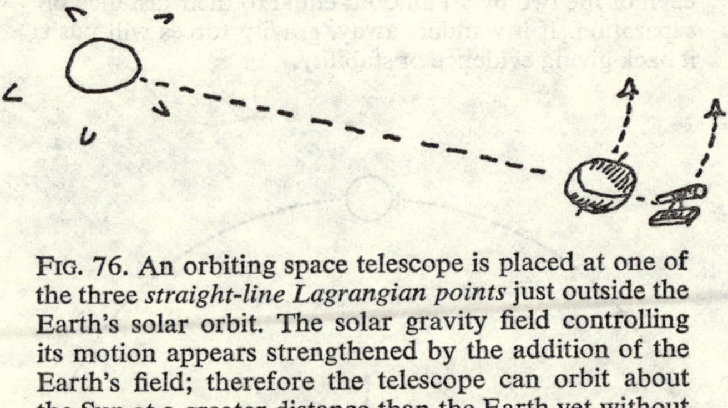

FIG. 76. An orbiting space telescope is placed at one of the three *straight-line Lagrangian points* just outside the Earth's solar orbit. The solar gravity field controlling its motion appears strengthened by the addition of the Earth's field; therefore the telescope can orbit about the Sun at a greater distance than the Earth yet without its period increasing beyond one year. By permanently lying within the Earth's shadow it is not subject to the thermal stresses which plague all optical instruments.

CHAPTER EIGHT

Notes from a Lonely Planet

Oriesa

". . . beep . . . beep beep . . . beep . . . beep beep . . . beep . . . beep beep . . ." The Director of the Drake–Sagan Cyclops Search Programme, Southern Hemispheric Division, was called in from his home in Alice Springs at three o'clock in the morning. There was no doubt about it. They were monitoring a pulse-modulated signal, at precisely 1.665 gigahertz, that is, 1665 million cycles (megacycles) per second. This was the frequency characteristic of the dissociation of water.

Throughout the 1980's the frustration had been unbearable. True enough, in the 1970's no one really expected success. Some of the studies paralleled the pioneering work of Frank Drake in 1960, where specific stars–in his case Epsilon Eridani and Tau Ceti, each 12 light years away–were examined only at the 1.42 gigahertz radiation emitted by neutral hydrogen. Others, such as Dixon and Cole in 1973, searched the Galaxy in all directions at that single frequency. Still others, such as Sagan and Drake from 1975 into the 1980's, concentrated their efforts on radio emissions at many frequencies from nearby galaxies, with the great Arecibo spherical-sectioned radiotelescope, 1000 feet across, set across the landscape as if it were a lifenet for giants from another world. This colossal instrument in Puerto Rico was duplicated in slightly larger versions in two more parts of the world by the close of the 1980's. The Cyclops array, on the other hand, had a greater collecting surface and was more versatile and in some ways easier to construct. Mimicking the eye of a fly, it collected 1500 separate images, radio images in this case, which were then computer processed. The Australian Cyclops, the first of a series to be built in the world, was in operation for barely two months in 1994, when the one–two signal, as it was called, was received and echoed around the world. It was the beginning of a complex pattern of messages, entirely pulse modulated and built up from a very simple structure. The Cyclops computer was able to decode the early messages, but as time went on a point was reached where the information seemed to extend far beyond the capability of any computer available on the Earth. The messages continued to pour in for a total of nine months after which time the cycle was repeated, starting again with the same simple pattern. By the fourth repetition it was apparent that the signal strength had gone down, evidently because other stellar systems were now being contacted. But of course there was no way to test this hypothesis.

The discovery of extraterrestrial intelligence stirred every human being on the Earth who perceived the message, when translated into a local language, as directed personally to him or her alone. The stupendous events of modern history, such as the atomic bomb detonations and the manned lunar landings, palled by comparison. An international group was formed to decipher the complete message. The immediate impact was a fivefold increase in the SETI—*s*earch for *e*xtra*t*errestrial *i*ntelligence—budget. The planned Mars expedition was placed on course for a 1999 landing of five astronauts, three men and two women, all scengineers (scientist-engineers).

But the one–two signal came not from Mars, not

from any of the presumed ten billion planetary systems in the Galaxy, but from a source beyond it in a galaxy in the constellation Sculptor. Sculptor is one of the local group of about 20 galaxies to which the Great Galaxy of Andromeda (M31) as well as our own Milky Way belong. Like our own Galaxy, M31 is a giant spiral filled with dust and gas still condensing into stellar and planetary systems. But Sculptor had long ago completed the condensation process. Its stars, and presumably planets, are hundreds of millions, if not billions, of years older than those of a spiral. The chance of encountering, by random search, an advanced civilization during its limited lifetime on a planet or satellite is consequently enhanced in an elliptical galaxy such as Sculptor.

Also enhanced is the probability that a radio signal sent from inside that galaxy in our direction would actually be picked up by our antennas. For one thing the density of gas and dust in an elliptical galaxy is low. Also, in a direction normal to our *Galactic Plane* the density of matter within our Galaxy which could attenuate a radio signal is low. These facts combine in our favor, for Sculptor lies almost exactly on the South Galactic Pole. Despite its enormous distance Cyclops can target in on a billion potential sources of life at once. In such a field of view a small fraction will be habitable, of which a portion of these will be inhabited at any given time. A very small number of these will be thriving advanced technological civilizations during our own observing period (or a half-million years ago allowing for the radiation transit time). And if we are lucky, one of these may just happen to be sending radio waves of sufficient power in our direction to be detected above the radio noise of the Universe. Indeed, we were lucky.

It is more than likely that the source of the signal was directing its beam toward a neighboring stellar system and the Australian Cyclops antenna picked up the leakage, celestial eavesdropping so to speak. There was no doubt that the signal was intended as a means of establishing contact. The binary code of the pulsed signal traced out a series of 1/0 diagrams picturing life forms. The 1's and 0's filled a 31 × 41 matrix (each prime numbers) consisting of 1271 bits, or elementary units of information. The similarity with the concept proposed by Oliver and Drake in the 1970's was remarkable, but similarities throughout the Universe need come as no surprise.

It would be an extraordinary coincidence if life as *we* know it were to evolve under dissimilar conditions, or even if conditions were reasonably similar. But we believe that factors such as the elements, basic compounds such as water, gravitational, electromagnetic, and nuclear forces, and the laws of mechanics, thermodynamics, and the like, are the same for all components of, and all civilizations in, the Universe. Each world will, sooner or later, have its Einsteins and Maxwells, each world will discover fire, chemical and nuclear, to threaten its own destruction and set bounds upon its lifetime. A number system based on 10 may be unique, and perhaps bipeds with six appendages on two hands may choose 12 as their base. But a binary system based on two is a universal language. As for energy, a world circling close to its sun has more than enough to develop and maintain its civilization, which includes sending a radio-maser beam* across the Universe.

The complex matrix pictures which followed the initial one–two contact signal began by teaching a written language so that gradually there was less and less reliance on pictures and more on linear information. The most characteristic attribute of the life forms was the total absence of hostility among individual units

* A maser–*m*icrowave *a*mplification by *s*timulated *e*mission of *r*adiation–is similar to a laser that operates in the visible portion of the electromagnetic spectrum.

and among complex groupings. The establishment of larger and more complex groupings without deleterious effects on individual components was the primary function of life. It was a means to an end, an end which one strove to reach but perhaps never could reach. In completing this life function, in attaining the higher and higher levels of complexity, life for them was more meaningful and the meaning of life became clearer. What was this end, always in sight yet seemingly out of reach? Unity with WorldMind? These are mere words. The signal taught us much and we could learn many things, mostly dealing with the combining of matter and forces into new technologies, or with biological forms and social structures. Weapons were without meaning. (The gulf between our two worlds far exceeded the physical space between them.) No matter now hard we tried, the messages in the final month could not be deciphered. It left humankind bewildered and uneasy as the signal began to fade and move on, perhaps to more receptive worlds.

This development of linkages, from element to element and group to group, was *synergistic,* that is, the whole was greater than the sum of its parts. As the structures became more complex the best was brought out and the undesirable discarded. Nevertheless, individualism was protected by instinct, to make use of those desirable fluctuations by which society advances, and to prevent social decay. The avoidance of conflict within and among groups was also an instinct, one as natural as eating.

There were so many questions we would have liked to have asked. Did life on this planet develop initially by Darwinian Natural Selection? Was this civilization the next stage beyond? Was this how nuclear holocaust was avoided? The Earth was breathless with the daily arrival of new signals. New technologies, new propulsion systems, new communication devices were one thing. But a non-Darwinian world based essentially on love and unrelated to struggle? This no one expected.†

The source of the radio signal, we were told, was the planet of a star near the edge of the galaxy. The density of stellar systems was low enough to require a high power, highly directional signal to reach neighboring stars, accounting for Cyclops' ability to receive it. The planet was half again as large as Earth and between the distance of Venus and Earth to the center of its star. The star was somewhat smaller than the Sun but a spectral type-G star as the Sun and Alpha Centauri. Water covered much of the planet's surface and its overall temperature was higher than Earth so polar caps were absent.

Usable energy was in abundance, all indirectly derived from its sun in electrical form. This was due in no small way to the planet's sociopolitical structure, a direct consequence of its "non-Darwinian" biostructure. Phrases such as "security of supply," in the sense of non-interference by life forms, were without significance in a world that knows no hostility. As on Earth its solar energy was stored in the wind; its wind energy was stored in the sea waves. Rainfall, also an intermittent form of solar energy, was stored in recesses behind mountains and dams. Wind-, wave-, and solar-electric generators were located in the best possible places–out on the sea for the first two or in deserts for the last–where the energy input was greatest and where few "people" would have to look at them. The entire global electric system was interconnected by undersea and underground cables, then sent to population centers. If the wind or waves suddenly

† They were non-Darwinian only in the sense of not evolving through competitive struggle. It is likely that Darwin would have maintained that the organism as a whole–and in this regard "organism" means the synergistic aggregate of all the inhabitants –formed this optimal way of adjusting to its surroundings, and survived. On Earth some primitive forms–the ants and bees– move in this direction. Humans try, but instinct usually wins out.

stopped, or a dark cloud passed over the desert, the multiple hydroelectric generators at the water-storage dams, always on automatic standby, would be ready to take over, in a fraction of a second, at almost any required power level. Thus the system needed none of the costly energy storage facilities in use on Earth such as batteries or flywheels which give back so little of the energy they take in. The peak energy needs of the daytime side of the planet came from the wind and wave generators at the twilight and night zones. They discovered a plastic dielectric, somewhat advanced over Earth's, which permitted direct-current cables with small-diameter aluminum cores to be charged to tremendous voltages. This enabled a cable to transmit several thousand megawatts with only a trickle of current, so that power losses in the cable were less than 1% per 10,000 kilometers. The cables lasted from 50 to 100 years, about the same as those on Earth. The wave generators proved to be the most reliable in terms of regularity of output. In their most active oceans there might be periods of calm for as long as four or five days, somewhat longer than in our North Atlantic, and the average output was 50 megawatts per kilometer, some 30% less than ours. But when one sea was calm another was active, or the solar cells or wind generators were active. Their computers showed that even a 10% global power failure had a probability several times smaller than their planet being destroyed by collision with a celestial intruder. This production of electricity, more than was needed, made it worthwhile to operate nearly all types of transportation and heating systems electrically. This included heat for warmth and heat for producing things. Pollution removal was never described in the Oriesa signal, for one cannot remove something that never existed in the first place.

All this was typical of the kinds of technical information which flowed to Earth. The energy system, so straightforward and sensible for Oriesa, stunned every nation on Earth. It made each realize that a safe, dependable, abundant, pollution-free, permanent energy system was technically and economically feasible by an entirely new approach. To get it depended on one thing alone: that each nation—the temporarily energy-rich and the permanently energy-poor—now and for all time must cooperate completely with all others. It was the price for survival.

The information continued to pour in—technology, life forms, social structure. People everywhere were asking the same questions. Was there some "Power" in the Universe that was behind it? And why especially now, here and now? Theologians and educators, scientists and statesmen, everyone pondered—*why?*

No one will ever know. It was only one planet, just beyond our Galaxy. One planet out of countless billions upon billions. Reaching out through space and time to form the ultimate grouping, couplings with worlds beyond itself.

A Universe teeming with life?

". . . Other worlds do not live. We will stir them with our limbs. Other parts of the Universe cannot see. We will bring them a gift of eyes. Where all is silence, this thing that we call human will speak." That was how Ray Bradbury put it. Others, like him, find many valid reasons for our colonizing of space. Still others find tantalizing the prospect that many advanced civilizations are already likely to have established colonies throughout their galaxies, or ours. This alone multiplies the chances of our finding life, or a signal from life, extended ("seeded" is a more appropriate word) throughout the Universe. They look at it as instinctive, this urge to explore, expand, and colonize. And then a simple extrapolation of this tendency to all advanced, technological (assuming the two words go together) civilizations, and the Universe teems with life.

But what are the motivations for human beings, Earthlings, to establish colonies? In many ways they are more pertinent now than during the Great Age of Colonization. An instinct for adventure is one thing, but in the long run it may be overshadowed by the instinct for survival. In the past one might be fleeing from political oppression, or from persistent famine, or searching for raw materials, or new markets.

But why now? In 1910 the Earth's population was a comfortable one billion. Since then it has quadrupled. But the waste products of a civilization that supports its present population, at a far higher per capita living standard, have quadrupled many times over. With the expansion of pollution of the air, sea, and land, society soon will exhaust its ability to cope. In Tokyo, for instance, the heat generated through energy use is a direct contributor to the microclimate. On a hot summer day in New York one often wonders to what extent all those hundreds of thousands of air conditioners are actually creating the climate that they are trying to avoid. Oil reserves are rapidly being depleted. There is still much coal left but its increased use is becoming a universal hazard. Strip mining defiles the land and uproots population centers. Power-plant combustion brings down sulphuric acid rain over a radius of a thousand kilometers or more. Microparticulates containing toxic trace metals from coal combustion and other sources pour by the ton through the fine filters at the smoke stacks, float over the cities and countryside as an almost invisible haze, gradually settle down on the streets and roadways and, bypassing the human body filter, impact onto lung tissue to remain there or be later translocated to other parts of the body.*

* The most poisonous elements found in coal, cadmium, arsenic, and antimony, volatilize at combustion temperatures and become adsorbed on microparticulates below about 3 microm-

Growth in population and parallel economic growth, even at only a couple of percent a year, exact their annual toll as "quality of life" becomes less and less equitable with "standard of living." One often hears that one thing or another will "solve" the major world problems. In agriculture it was DDT and later the Green Revolution. In education, manufacturing, and distribution it was the electronic computer. In energy supply it was the nuclear fusion reactor. These can solve pressing problems, at least for a time. But each year the net population *increase* of Earth equals the present population of the United Kingdom plus Ireland and Denmark.†

What we would like to find is a way out. One which can eventually relieve population pressures and its by-products and the drain on limited supplies of raw materials. Space colonization offers at least a possibility, a feasible way, to do this. A superficial look at space colonization would cause it to be dismissed at once. Have we not enough problems here on the Earth without taking on new ones? Why drain off our dwindling resources in a colossal project which dwarfs by far even Apollo? The list goes on.

But a *careful* look demands that we be practical. Basically we need ask only two questions: 1) At what level and in what time scale can a space colony be self-sustaining? 2) At what level and in what time scale can

---

eters in diameter, the respirable range. The fraction adsorbed increases markedly with the reciprocal of the radius, for a small particle presents a larger surface per unit volume than a large one. (Surface area is proportional to the square of the radius of a sphere, whereas volume is proportional to the cube of the radius.)

† It is interesting to note that at the current rate of energy usage, there is enough deuterium in the oceans to supply the hypothetical fusion reactors and satisfy the world's energy appetite for twenty *billion* years. However, if energy demand were to increase steadily by a mere 1% per year, the deuterium supply would be used up in less than 1850 years!

a space colony produce a surplus and return the investment back to the Earth? The assumption in both cases is that the investment in labor, energy, and materials is initially available on Earth as a small fraction of its total resource. As for labor, the unemployment of tens of millions of workers in the industrial world alone shows no signs of diminishing as we move into the 1980's.

I recall a radio interview with Arthur C. Clarke on the day the Eagle landed and Neal Armstrong was about to test his footing in the lunar dust. Up to that moment the astronautical cost was astronomical. However, Clarke pointed out, if you compare the total cost of Apollo with the total public expenditure for cosmetics, or the amount drained off by the Mafia, we get an entirely different perspective. Many people in the United States regarded the whole space program as a) the all (prime-) time TV Spectacular, b) a real leap forward for the science of geology (sic!), and c) a chance to get back at the Russians for orbiting the first Sputnik. Perhaps some still see it only in those terms and wonder why the funding continues into the 1980's.

The most important results of the manned and unmanned space program can be realistically summed up: a) An increase in scientific knowledge in interrelated disciplines (planetary science, biology, relativity, etc.), b) technological development with emphasis on reliability (solar-electric generation, rocket propulsion, etc.), and c) a gain of experience in all aspects of space activities (effects on life systems in zero gravity, repairing and maintaining equipment, etc.). These might be the list of a pragmatic economist. A theologian, philosopher, or cosmologist might come up with a different set of results.

The economist might ask for a cost-benefit analysis of space activities to date. He would say that solar cells would have come anyway, though at a slower pace; likewise for computers. He would certainly deny that the space adventure of the 1960's played any significant role in the long minidepression of the 1970's; the world munitions tab is some 20 times greater. He would not count much for the economic value of prestige, but he would grant some points to the contributions toward basic science which eventually show up as positive gains on the economic ledger. In weighing the overall benefits against the total cost he would ask the one pertinent question above all others: Where do we go from here?

When most people think of space colonies they naturally think along the lines of expeditions to Mars, or even Venus, certainly the Moon, and possibly, if not Jupiter, then one or two of its 14 moons, those covered with water–ice. The four large moons of Jupiter, Galilean moons as they are called, are the size of small planets; Titan, the largest of Saturn's ten moons, is the largest moon in the Solar System. Titan manages to retain solar heat near its surface by means of its atmosphere which is rich in those triatomic gases, such as water vapor, which absorb and re-radiate strongly in the infrared region of the electromagnetic spectrum. Thus, while visible light from the Sun actively heats up the surface, it radiates at the wavelength region which corresponds to that temperature, namely infrared. Instead of losing this radiation to space, however, the atmospheric molecules will be excited and will reradiate it in all directions, so that half of it, at least, finds its way back to the surface, heating it further. This phenomenon, erroneously called the *Greenhouse Effect,*‡ also takes place on Venus, where it, together with violent duststorms, heats up the surface to 700°C, eliminating all chances of finding life in forms with which we are familiar.

‡ The Greenhouse Effect is a blocking, by a physical barrier such as glass or plastic film, of rising warm air. The contribution toward heating by infrared re-radiation of glass is minor, and does not exist at all for the film.

On these planets and satellites we can construct shell-like structures retaining some semblance of a hospitable environment, grow food in the soil with the aid of nutrients from Earth, and mine the surface for constructional materials. Some scientists have suggested that through *planetary engineering* techniques, it would be possible to alter the Martian environment to provide the necessary oxygen, water, moderate temperature, and ultraviolet protection to at least approach conditions on Earth. To generate oxygen, for instance, one can bring in colonies of microorganisms which can withstand Martian conditions, especially the radiation flux. Through photosynthesis, their expired oxygen will fill the atmosphere (in the way that plant life on Earth, for millions of years, has been responsible for our oxygen). To accelerate the process and make the planet more comfortable, spreading dark material on the icy surface will reduce the planet's reflectivity, its *albedo*. This will vaporize the polar caps gradually, but eventually it will raise the temperature by a reasonable amount.

Apart from the Moon, possible colonizing regions take much too long to reach. Mars takes more than a half year; voyages to the Jovian and Saturnian moons are far longer. As for the Moon, it suffers from several disadvantages, not the least of which is the 14-day night. Its gravity, on the one hand, is too low for comfortable living for long periods, yet, on the other hand, it is just high enough to be a nuisance when transporting goods and people back and forth to and from Earth.

A made-to-order orbiting world

Suppose we wanted to design our own space colony with all the comforts of home, or at least as many as we can provide, and as few of the discomforts as possible. What would we need and where would we place it?

Living space, gravity, radiation protection, sunlight on and off, water, food, ample energy, trout streams, books, plays, etc. The list could be longer of course, depending on individual needs. It may be difficult to produce a weekend at Mount Snow or a summer at Costa del Sol; anyway, duplicates are rarely as good as the real thing. But there are compensations, many compensations.

First, where would we place it? A low-orbiting Earth satellite is out of the question, not only because it may encounter small amounts of atmospheric drag, or tidal forces, but because it will move in and out of the Earth's shadow every hour or two. This will be not only a discomfort but an inconvenience with respect to the stabilizing of the instrumentation and equipment, which control the structure and regulate the climate, for instance.

A colony in a geosynchronous orbit fares slightly better in these respects. But to see the problem in sharper perspective, we want to be independent of the Earth and take full advantage of the Moon and eventually asteroids. What better place to orbit the colony than in one of the Earth–Moon equilateral-triangle Lagrangian points? It would sit in a stable gravitational pocket equidistant from the Earth and Moon and therefore could be found from the Earth in the Moon's orbit but far from the Moon. The energy to get people and goods to the colony from Earth is no more than to the Moon, in fact it would be less since the colony has no gravity to overcome. Yet it rotates just enough to produce 1 g_e at the curved inner surface. Docking takes place at the hub, where gravity is zero. We saw in Chapter Two that the energy cost, based on nighttime supply at the gate of an electric power plant, in moving from the Earth to "infinity" was but 21 cents a pound (it would come to much more if we threw away the power plant after each run). Even if we relied on chemical propulsion, the *space shuttle* will meet us halfway in lowering costs.

Electric propulsion could become a reality, even today, for sending small bucket payloads of lunar material to the colony at the Lagrangian point. By means of magnetic levitation, the TLA, or *transport linear accelerator*, holds the bucket, containing superconducting coils, over a "linear synchronous motor track" that flings it toward the Lagrangian point. As the lunar gravity field weakens, the payload hurtles through space at a slower and slower speed, eventually matching the speed of the colony as well as the Moon, all three of which move about the Earth together. The energy needed for lunar escape is but 5% of Earth escape. At the current price for electricity this would be but a penny per pound! It would be fairer to calculate the capital cost of the parabolic solar collector, the steam-turbine generator, and the TLA equipment, and take into consideration the volume of material sent and the maintenance required. Perhaps the maintenance costs will be low (though I would guess the hourly wages would be high) and as the sunlight is free we might bring the costs down to less than a cent a pound. I would hope so, but I would not bet on it.

What is the nature of the payload? 40% of the Moon's surface is oxygen. Mixed with 12% of hydrogen by weight (in liquid form from the Earth), we then have ample water; unmixed it is an excellent rocket propellant. Then there is iron, which will be fashioned into steel for structural members, including, along with residual slag from manufacturing operations, material to serve as a cosmic-ray shield for the colony. There would be silicon, from which solar cells will be fashioned, manganese, chrome, and last but not least, aluminum. To produce aluminum from ore is especially energy intensive and energy is the one thing in greater supply than all others for all human activities at the space colony.

Princeton physicist Gerard K. O'Neill, in a memorable article in *Physics Today,** showed how space colonies could be built with minimal dependence on the Earth, and produce offspring with even less dependence. All this could come out of raw materials from the Moon, and eventually from asteroids (iron, nickel, cobalt, copper, etc.). Steel factories, aluminum factories, electronic factories. All operating near the rotating axis at low- and zero-gravity conditions, in pure vacuum, under controlled temperatures. Solar process heat and steam-boiler produced electric power from huge barely supported parabolic mirrors unstressed by tidal forces. An alternation of day and night with movable flat mirrors. Centrifugal rotation to simulate gravity. And houses, trees, rolling hills, meadows, farm animals, lakes, streams, clouds. A simulation of Earth itself but with one major difference–no pollution.

And with the metallurgical–electronic laboratories and lunar silicon, the long awaited solar cell (photovoltaic) electric generator and microwave converter is fabricated and towed to the geosynchronous orbit, separated from the Lagrangian point by only 2.1 kilometers/second. Its design, unconstrained by the launch vehicle, is simpler and cheaper to manufacture, like most of the colony's products. It becomes one of many dividends on the Earth's investment.

* * *

This book has moved in two worlds, the World of Reality and the World of Fantasy. How can we keep them apart?

It does not seem so long ago. It was called "Futurama," the major exhibit of a World's Fair that described itself as "The World of Tomorrow." The fabulous world of 1960 unfolding before millions of wide-eyed onlookers. Highways resembling velvet rib-

* Gerard K. O'Neill, *Physics Today,* September 1974, pp. 32–40. The concepts were amplified in his book, *The High Frontier* (New York: William Morrow and Co., 1977).

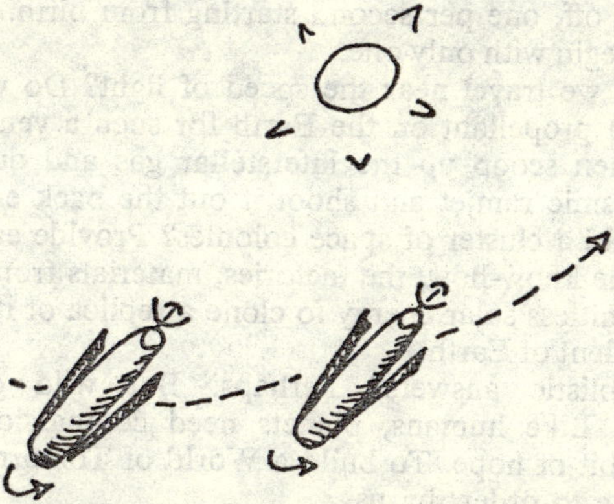

FIG. 77. A long cylindrical space colony with flat mirrors for reflecting sunlight to the interior, and parabolic mirrors for concentrating solar energy, to provide industrial process heat and steam-turbine-produced electrical energy. The cylinder rotates axially to simulate gravity in the interior; however its angular momentum will keep it pointing at a star rather than the Sun as it revolves with the Earth–Moon system around the Sun. To make it easy to aim at the Sun, rotating-cylinder colonies should be paired, each turning in opposite directions so that the net angular momentum is zero.

bons gliding by orchards of glass-enclosed fruit trees and gleaming white futuristic cities, then leaping across clear rivers in graceful suspended arches five levels deep. It did not turn out quite that way, but seen from 1939 almost anything was possible.

Before the 1960's were over, "one giant step . . ." had been taken for every human being on Earth. The opportunities into the next century and beyond are almost without limit. There may indeed be ten billion life-supporting stellar-planetary systems in our Galaxy, a number so overwhelming that the High Lama of Shangri-La would have to live 300 years to count each system off, one per second starting from birth. But we need begin with only one.

Can we travel near the speed of light? Do we have enough propellant on the Earth for such a venture? If not, then scoop up the interstellar gas and dust with our cosmic ramjet and shoot it out the back end. Can we build a cluster of space colonies? Provide each unit with the know-how, the factories, materials from space, and limitless solar energy to clone a replica of itself, independent of Earth.

Simplistic answers? Perhaps. But wild dreams? Never! Like humans, planets need compassion. And just a bit of hope. To build a World of Tomorrow that is made to order, by us.

Index